Ben Stacy Jerrik (Ed.)

Wittenberge

Ben Stacy Jerrik (Ed.)

Wittenberge

Municipalities of Germany, Regierungsbezirk, Gemeinde, Miesbach, Bad Wiessee

Part Press

Imprint

Permission is granted to copy, distribute and/or modify this document under the terms of the GNU Free Documentation License, Version 1.2 or any later version published by the Free Software Foundation; with no Invariant Sections, with the Front-Cover Texts, and with the Back- Cover Texts. A copy of the license is included in the section entitled "GNU Free Documentation License".

All parts of this book are extracted from Wikipedia, the free encyclopedia (www.wikipedia.org).

You can get detailed informations about the authors of this collection of articles at the end of this book. The editors (Ed.) of this book are no authors. They have not modified or extended the original texts.

Pictures published in this book can be under different licences than the GNU Free Documentation License. You can get detailed informations about the authors and licences of pictures at the end of this book.

The content of this book was generated collaboratively by volunteers. Please be advised that nothing found here has necessarily been reviewed by people with the expertise required to provide you with complete, accurate or reliable information. Some information in this book maybe misleading or wrong. The Publisher does not guarantee the validity of the information found here. If you need specific advice (f.e. in fields of medical, legal, financial, or risk management questions) please contact a professional who is licensed or knowledgeable in that area.

Any brand names and product names mentioned in this book are subject to trademark, brand or patent protection and are trademarks or registered trademarks of their respective holders. The use of brand names, product names, common names, trade names, product descriptions etc. even without a particular marking in this works is in no way to be construed to mean that such names may be regarded as unrestricted in respect of trademark and brand protection legislation and could thus be used by anyone.

Cover image: www.ingimage.com
Concerning the licence of the cover image please contact ingimage.

Publisher:
Part Press is a trademark of
International Book Market Service Ltd., 17 Rue Meldrum, Beau Bassin, 1713-01 Mauritius
Email: info@bookmarketservice.com
Website: www.bookmarketservice.com

Published in 2012

Printed in: U.S.A., U.K., Germany. This book was not produced in Mauritius.

ISBN: 978-613-8-69584-4

Contents

Articles

References

Wittenberge

Wittenberge	
 Wittenberge	
Coordinates	53°00′N 11°45′E
Administration	
Country	Germany
State	Brandenburg
District	Prignitz
Town subdivisions	7 Stadtteile/Stadtbezirke
Mayor	Oliver Hermann (—)
Basic statistics	
Area	50.44 km^2 (**unknown operator: u'strong'** sq mi)
Elevation	25 m (82 ft)
Population	18345 *(31 July 2011)*
- Density	364 /km^2 (942 /sq mi)
Other information	
Time zone	CET/CEST (UTC+1/+2)
Licence plate	PR

Postal code	19322
Area code	03877
Website	www.wittenberge.de [1]

Wittenberge (German pronunciation: [ˌvɪtənˈbɛɐ̯gə]) is a town of eighteen thousand people on the middle Elbe in the district of Prignitz, Brandenburg, Germany.

Geography

Wittenberge is situated at the right (north-eastern) bank of the middle Elbe at its confluence with the Stepenitz and Karthane in the German district of Prignitz. Within the same district, the town borders the *Ämter* Lenzen-Elbtalaue and Bad Wilsnack/Weisen as well as the district capital Perleberg. The *Verbandsgemeinde* Seehausen (Altmark) in the district of Stendal, Saxony-Anhalt, lays on the opposite side of the Elbe.

History

The Rathaus building

The site was marked out in 1239 at Wendischwalde and in 1300 the Saxon king, Otto I, established the settlement. It grew slowly but steadily. The town castle (1669) survives as the town museum but it suffered fires in 1686 and 1757, and floods from the Elbe in 1709 and 1761. Nevertheless, because of its position, it prospered from the opening of the Berlin–Hamburg steamship traffic. By 1823 milling, processing and refining of foodstuffs were important industries. In 1846 the railway came and a silk factory was built, followed by a chemical factory in 1849. The Reichsbahn railway workshops were established in 1875 and the world-famous Singer Sewing Machine Company set up there in 1903. By 1914 the Rathaus was built.

From August 15, 1942 until February 17, 1945 a concentration camp was established in Wittenberge. The camp was a subcamp to the Neuengamme concentration camp.[2]

Between the wars much architecture characteristic of the period was built in the form of workers' housing for the rapidly increased population. It suffered as most German towns did during WW II. Only the railway workshops remain as significant employers of the old industries but the large, skilled workforce remains for the newer industries of precision engineering and applied chemistry.

Notable residents

- Klaus Havenstein (1922 – 1998), actor
- Richard Z. Kruspe (b. 1967), musician (Rammstein)

References

[1] http://www.wittenberge.de/

[2] The camp is listed as No. 1587 Wittenberge, Kreis Wittenberge in the official German list of concentration camps Verzeichnis der Konzentrationslager und ihrer Außenkommandos (http://bundesrecht.juris.de/begdv_6/anlage_6.html) (German)

Westerrönfeld

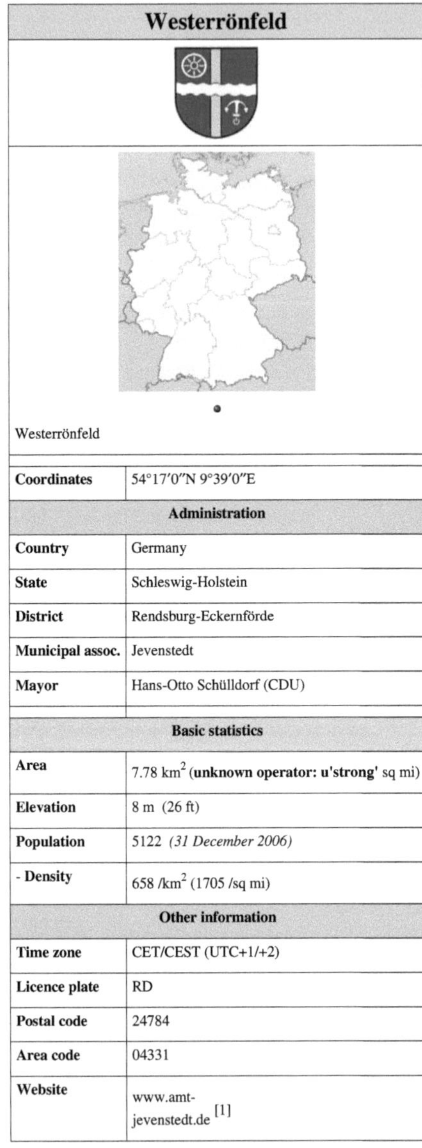

Westerrönfeld	
Westerrönfeld	
Coordinates	54°17′0″N 9°39′0″E
Administration	
Country	Germany
State	Schleswig-Holstein
District	Rendsburg-Eckernförde
Municipal assoc.	Jevenstedt
Mayor	Hans-Otto Schülldorf (CDU)
Basic statistics	
Area	7.78 km^2 (**unknown operator: u'strong'** sq mi)
Elevation	8 m (26 ft)
Population	5122 *(31 December 2006)*
- **Density**	658 /km^2 (1705 /sq mi)
Other information	
Time zone	CET/CEST (UTC+1/+2)
Licence plate	RD
Postal code	24784
Area code	04331
Website	www.amt-jevenstedt.de [1]

Westerrönfeld is a municipality in the district of Rendsburg-Eckernförde, in Schleswig-Holstein, Germany.

References

[1] http://www.amt-jevenstedt.de/

Municipalities of Germany

Municipalities (*Gemeinde*) are the official lowest level of territorial division **in Germany**. This may be the fourth level of territorial division in Germany, apart from those states which include *Regierungsbezirke* (singular: *Regierungsbezirk*), where municipalities then become the fifth level.

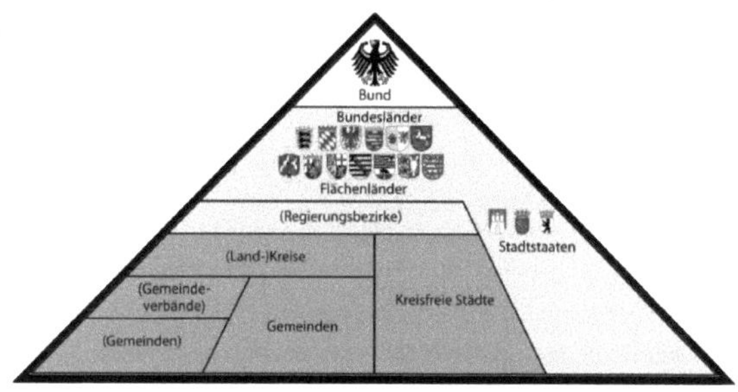

The "pyramid" of German administrative subdivisions

Overview

With more than 3,400,000 inhabitants, the most populated one is the city of Berlin; and the least populated is Wiedenborstel (8 inhabitants in 2010), in Schleswig-Holstein

Municipalities per federal state

List updated at August 1, 2009.

Federal state	Municipalities	M. with city status	Average nr. of inhabitants[1]	Average surface (km²)[2]	List (Cities, Towns, Municipalities)
Baden-Württemberg	1,101	312	9,764	32.41	C, T, M
Bavaria	2,056	315	6,090	33.03	C, T, M
Berlin	1	1	3,416,255	891.02	Berlin
Brandenburg	419	112	6,052	70.36	C, T, M
Bremen	2	2	331,541	202.14	Bremen, Bremerhaven
Hamburg	1	1	1,770,629	755.16	Hamburg
Hesse	426	189	14,255	48.80	C, T, M
Lower Saxony	1,022	164	7,800	45.25	C, T, M
Mecklenburg-Vorpommern	818	84	2,053	28.34	C, T, M
North Rhine-Westphalia	396	268	45,446	86.08	C, T, M
Rhineland-Palatinate	2,306	123	1,754	8.61	C, T, M
Saarland	52	17	19,935	49.40	C, T, M
Saxony	491	178	8,595	37.51	C, T, M
Saxony-Anhalt	851	118	2,835	24.03	C, T, M
Schleswig-Holstein	1,116	63	2,542	14.07	C, T, M
Thuringia	955	126	2,397	16.93	C, T, M

Germany	12,013	2,073	6,844	29.35	C, T, M

References

[1] Source: Statistisches Bundesamt (as of December 31, 2007)
[2] Source: Statistisches Bundesamt (as of December 31, 2006), for the computation less the municipality-free areas

See also

- List of municipalities in Germany

Gemeinde

Gemeinde (plural: ***Gemeinden***) is a German word for borough, commune, community, township, municipality, or in religious contexts, a parish or congregation (sometimes, *Kirchengemeinde*). [1]

Gemeinde also means, approximately, the parish or the community. This term in its theological usage, means more than just a community or parish. It has no English equivalent except among Mormons, who call it a ward. It encompasses the meaning of parish assembly (as a group entity) but also the congregation as a *koinonia* (Greek) or fellowship of believers.

See also

- Gemeente
- Gmina
- Municipalities of Germany
- Municipalities of Austria
- Communes of Switzerland
- Municipalities of South Tyrol

Notes

[1] "Travlang's German-English On-line Dictionary" (result for "Gemeinde"), Travlang (travlang.com), 2000, webpage: travlang-Gemeinde (http://dictionaries.travlang.com/GermanEnglish/dict.cgi?query=Gemeinde&max=50).

Regierungsbezirk

In Germany, a **Government District**, in German: ***Regierungsbezirk*** – pronounced [ʁeˈɡiːʁʊŋsbəˌtsɪʁk] is a subdivision of certain federal states (*Bundesländer*).

They are above the *Kreise*, *Landkreise*, and *kreisfreie Städte*.[1] The *Regierungsbezirk* is governed by a *Bezirksregierung* (literally "district government") and led by a *Regierungspräsident* (literally "government president").

Regierungsbezirke as from 1st of August 2008. In the map there are also shown the former RBs of Lower Saxony, Rhineland-Palatinate and Saxony-Anhalt

History

The first *Regierungsbezirke* were created by the Kingdom of Prussia, which divided its provinces into 25 *Regierungsbezirke* in 1808/1816. The *Regierungsbezirke* of North Rhine-Westphalia are in direct continuation of those created in 1815. Other states of the German Empire created similar entities, named *Kreishauptmannschaft* (in Saxony) or *Kreis* (in Bavaria and Württemberg) (not to be confused with the *Kreis* or *Landkreis* today). During the Third Reich, the Nazi government unified the naming; since then all these entities are called *Regierungsbezirk*.

On January 1, 2000 Rhineland-Palatinate disbanded its three *Regierungsbezirke* Koblenz, Rheinhessen-Pfalz and Trier - the employees and assets of the three *Bezirksregierungen* were converted into three public authorities responsible for the whole state, each covering a part of the former responsibilities of the *Bezirksregierung*.

On January 1, 2004, Saxony-Anhalt disbanded its three *Regierungsbezirke*: Dessau, Halle and Magdeburg. The responsibilities are now covered by a *Landesverwaltungsamt* with three offices at the former seats of the *Bezirksregierungen*.

At the foundation of Lower Saxony in 1946 by the merger of the three former Free States of Brunswick, Oldenburg, Schaumburg-Lippe and the former Prussian province of Hanover the former two states became *Verwaltungsbezirke* (roughly administrative regions of extended competence) within Lower Saxony besides the less autonomous Prussian-style *Regierungsbezirke* comprising the former Province of Hanover and the tiny Schaumburg-Lippe. These differences were levelled on 1 January 1978, when four territorially redeployed Regierungsbezirke replaced the two Verwaltungsbezirke and the old six Regierungsbezirke: Brunswick and Oldenburg as well as Aurich, Hanover (remaining mostly the same), Hildesheim, Lüneburg (old), Osnabrück and Stade. On January 1, 2005, Lower Saxony disbanded its remaining four Regierungsbezirke: Braunschweig, Hanover, Lüneburg, and Weser-Ems.

In 2005, North Rhine-Westphalia planned to abolish its five *Regierungsbezirke* and create three self-government entities. The old, "Prussian-style", Regierungsbezirk had no self-government organs.

On August 1, 2008, Saxony restructured its districts (*Landkreise*) and changed the name of its *Regierungsbezirke* to *Direktionsbezirke*. This was necessary because one of the new districts did not fit with the borders of the old

Regierungsbezirke and some responsibilities are now covered by the districts. The *Direktionsbezirke* are still named *Chemnitz*, *Dresden* and *Leipzig*. The authorities of the *Direktionsbezirke* are named *Landesdirektion* and their Presidents are called *Präsident der Landesdirektion* instead of *Regierungspräsidium* and *Regierungspräsident*.

Local existence

Not all *Bundesländer* have this subdivision; some are directly divided into districts. Currently, five states are divided into 22 *Regierungsbezirke*, ranging in population from 5,255,000 (Düsseldorf) to 1,065,000 (Gießen):

- Baden-Württemberg: Freiburg, Karlsruhe, Stuttgart, Tübingen
- Bavaria: Upper Bavaria, Lower Bavaria, Upper Palatinate, Upper Franconia, Middle Franconia, Lower Franconia, Swabia
- Hesse: Darmstadt, Gießen, Kassel
- North Rhine-Westphalia: Arnsberg, Cologne, Detmold, Düsseldorf, Münster
- Saxony: Chemnitz, Dresden, Leipzig

The regional governments of the *Regierungsbezirke* are mostly concerned with administrative decisions on a regional level. Most administrative routines are handled by the municipal government (of the city or the *Kreis*) while legislation is passed by the parliament of the *Bundesland* or on a national level.

References

[1] Regional Governments in France, Germany, Poland and The Netherlands (HTML version of PowerPoint presentation) (http://google.com/search?q=cache:I1oTRcSOTJ8J:www.hanse-passage.net/hansepassage/hpcms/uploads/_presentation%20analysis.ppt+"regierungsbezirk+is"&hl=en&ct=clnk&cd=18&lr=lang_en) - Cachet, A (coordinator), Erasmus University, Rotterdam

Rendsburg-Eckernförde

Rendsburg-Eckernförde	
— District —	
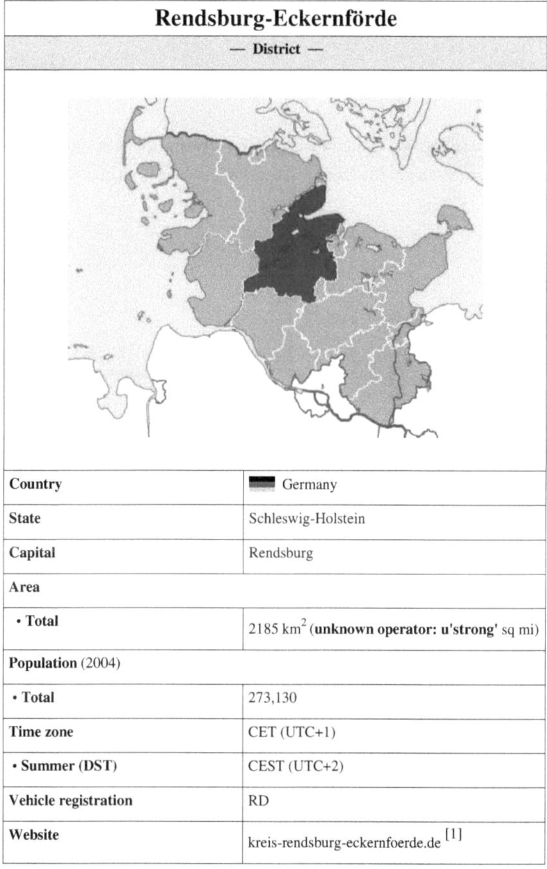	
Country	Germany
State	Schleswig-Holstein
Capital	Rendsburg
Area	
• **Total**	2185 km^2 (**unknown operator: u'strong'** sq mi)
Population (2004)	
• **Total**	273,130
Time zone	CET (UTC+1)
• **Summer (DST)**	CEST (UTC+2)
Vehicle registration	RD
Website	kreis-rendsburg-eckernfoerde.de [1]

Rendsburg-Eckernförde (Danish: *Rendsborg-Egernførde*) is a district in Schleswig-Holstein, Germany. It is bounded by (from the east and clockwise) the city of Kiel, the district of Plön, the city of Neumünster, the districts of Segeberg, Steinburg, Dithmarschen and Schleswig-Flensburg, and the Baltic Sea.

History

In 1867 the Prussian administration established twenty districts in its province of Schleswig-Holstein, among them the districts of Rendsburg and Eckernförde. The present district was established in 1970 by merging the former districts. Rendsburg was part of Denmark till the Prussian-Danish war. Prussia won and Rendsburg became part of it. In the years following Prussia became part of Germany and Rendsburg also.

Geography

The district is situated at the coast of the Baltic Sea, roughly between the cities of Schleswig and Kiel. A large portion of the Kiel Canal passes through Rendsburg-Eckernförde.It is one of the largest districts in the whole of Germany.

Coat of arms

The coat of arms displays:

- two lions (blue on yellow) from the arms of Schleswig
- a nettle leaf (white on red) from the arms of Holstein

Towns and municipalities

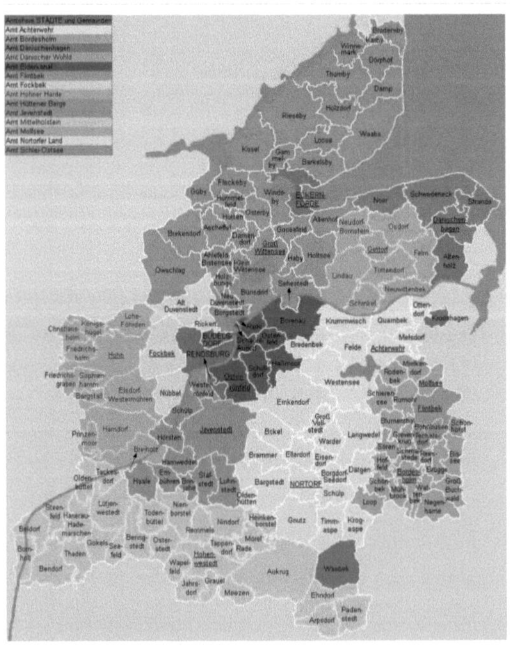

Independent towns and municipalities

- Büdelsdorf
- Eckernförde
- Rendsburg
- Altenholz
- Kronshagen
- Wasbek

Ämter

- **1. Achterwehr**
- Achterwehr[1]
- Bredenbek
- Felde
- Krummwisch
- Melsdorf
- Ottendorf
- Quarnbek
- Westensee
- **2. Bordesholm**
- Bissee
- Bordesholm[1]
- Brügge
- Grevenkrug
- Groß Buchwald
- Hoffeld
- Loop
- Mühbrook
- Negenharrie
- Reesdorf
- Schmalstede
- Schönbek
- Sören
- Wattenbek
- **3. Dänischenhagen**
- Dänischenhagen[1]
- Noer
- Schwedeneck
- Strande
- **4. Dänischer Wohld**
- Felm
- Gettorf[1]
- Lindau
- Neudorf-Bornstein
- Neuwittenbek
- **6. Flintbek**
- Böhnhusen
- Flintbek[1]
- Schönhorst
- Techelsdorf
- **7. Fockbek**
- Alt Duvenstedt
- Fockbek[1]
- Nübbel
- Rickert
- **8. Hohner Harde**
- Bargstall
- Breiholz
- Christiansholm
- Elsdorf-Westermühlen
- Friedrichsgraben
- Friedrichsholm
- Hamdorf
- Hohn[1]
- Königshügel
- Lohe-Föhrden
- Prinzenmoor
- Sophienhamm
- **9. Hüttener Berge**
- Ahlefeld-Bistensee
- Ascheffel
- Borgstedt
- Brekendorf
- Bünsdorf
- Damendorf
- Groß Wittensee[1]
- Haby
- Holtsee
- Holzbunge
- Hütten
- **10. Jevenstedt**
- Brinjahe
- Embühren
- Haale
- Hamweddel
- Hörsten
- Jevenstedt[1]
- Luhnstedt
- Schülp bei Rendsburg
- Stafstedt
- Westerrönfeld
- **11. Mittelholstein**
- Arpsdorf
- Aukrug
- Beldorf
- Bendorf
- Beringstedt
- Bornholt
- Ehndorf
- Gokels
- Grauel
- Hanerau-Hademarschen
- Heinkenborstel
- Hohenwestedt[1]
- Jahrsdorf
- Lütjenwestedt
- Oldenbüttel
- Meezen
- Mörel
- Nienborstel
- Nindorf
- Osterstedt
- Padenstedt
- Rade bei Hohenwestedt
- Remmels
- **12. Molfsee**
- Blumenthal
- Mielkendorf
- Molfsee[1]
- Rodenbek
- Rumohr
- Schierensee
- **13. Nortorfer Land**
- Bargstedt
- Bokel
- Borgdorf-Seedorf
- Brammer
- Dätgen
- Eisendorf
- Ellerdorf
- Emkendorf
- Gnutz
- Groß Vollstedt
- Krogaspe
- Langwedel
- Nortorf[1, 2]
- Oldenhütten
- Schülp bei Nortorf
- Timmaspe
- Warder
- **14. Schlei-Ostsee** [seat: Eckernförde]
- Altenhof
- Barkelsby
- Brodersby
- Damp
- Dörphof
- Fleckeby
- Gammelby
- Goosefeld
- Güby

- Osdorf
- Schinkel
- Tüttendorf
- **5. Eiderkanal**
- Bovenau
- Haßmoor
- Ostenfeld
- Osterrönfeld[1]
- Rade bei Rendsburg
- Schacht-Audorf
- Schülldorf

- Klein Wittensee
- Neu Duvenstedt
- Osterby
- Owschlag
- Sehestedt

- Seefeld
- Steenfeld
- Tackesdorf
- Tappendorf
- Thaden
- Todenbüttel
- Wapelfeld

- Holzdorf
- Hummelfeld
- Karby
- Kosel
- Loose
- Rieseby
- Thumby
- Waabs
- Windeby
- Winnemark

[1]seat of the Amt;[2]town

References

[1] http://www.kreis-rendsburg-eckernfoerde.de

External links

- Official website (http://www.kreis-rendsburg-eckernfoerde.de/) (German)

Amt (country subdivision)

Amt is a type of administrative division governing a group of municipalities, today only found in Germany, but formerly also common in northern European countries. Its size and functions differ by country and the term is roughly equivalent to a U.S. township or county.

Germany

Prevalence

The *Amt* (plural, *Ämter*) is unique to the German Bundesländer (federal states) of Schleswig-Holstein, Mecklenburg-Western Pomerania and Brandenburg. Other German states had this subdivision in the past. Some states have similar administrative units called Samtgemeinde (Lower Saxony), Verbandsgemeinde (Rhineland-Palatinate) or *Verwaltungsgemeinschaft* (Baden-Württemberg, Bavaria, Saxony, Saxony-Anhalt, Thuringia).

Definition

An *amt*, as well as the other above-mentioned units, is subordinate to a district *(Kreis)*, and is a collection of municipalities. The amt is lower than district-level government, but higher than municipal government, and may be described as a supra-municipality or "municipal confederation." Normally it consists of very small municipalities; larger municipalities do not belong to an *amt*, and are called *amtsfreie Gemeinden*(independent municipalities).

Former usage in other countries

Denmark

The *amt* (plural, *amter*; English, "County") was an administrative unit of Denmark (and, historically, of Denmark-Norway). The counties were established by royal decree in 1662 as replacements for the former *Len* (fiefs). The *amter* were originally composed of independent towns (*købstæder*) and parishes, and held only small areas of responsibility. During the 20th century, they were granted responsibility for the hospital service for the non-urban population. A 1970 administrative reform reduced the number of counties to fourteen and eliminated the administrative distinction between (rural) parish and town. From then on, the *amter* were composed of a number of municipalities (Danish: *kommuner*). The reform granted the counties wider areas of responsibility, most notably running the national health service and the *gymnasium* secondary schools.

The Danish Municipal Reform of January 1, 2007 abolished the *amter* and replaced them with five administrative regions, now solely charged with running the national health service. In contrast to the *amter* the regions hold no authority to levy taxes. The reform re-delegated all other areas of responsibility to either the municipalities or the state. At the same time, smaller municipalities were merged into larger units, cutting the number of municipalities from 270 to 98. See Counties of Denmark for more information about the Danish usage of the term.

Netherlands and Flanders

Ambacht can be seen as Dutch equivalent to *amt*. *Ambachten* existed in Holland, Zeeland and Flanders up to about 1800.

Norway

From 1662 to 1919, the counties of Norway were called *amter*. They are now referred to as *fylker*.

References

Schleswig-Holstein

Schleswig-Holstein	
— State of Germany —	
Flag	Coat of arms
Coordinates: 54°28′12″N 9°30′50″E	
Country	Germany
Capital	Kiel
Government	
• Minister-President	Peter Harry Carstensen (CDU)
• Governing parties	CDU / FDP
• Votes in Bundesrat	4 (of 69)
Area	
• Total	15763.18 km^2 (**unknown operator: u'strong'** sq mi)
Population (2007-09-30)[1]	
• Total	2837021
• Density	**unknown operator: u'strong'**/km^2 (**unknown operator: u'strong'**/sq mi)
Time zone	CET (UTC+1)
• Summer (DST)	CEST (UTC+2)
ISO 3166 code	DE-SH
Vehicle registration	formerly: S (1945–1947), SH (1947), BS (1948–1956)[2]
GDP/ Nominal	€ 75.63 billion (2010) [3]

NUTS Region	DEF
Website	schleswig-holstein.de [4]

Schleswig-Holstein (pronounced [ˈʃleːsvɪç ˈhɔlʃtaɪn] (listen)) is the northernmost of the sixteen states of Germany, comprising most of the historical duchy of Holstein and the southern part of the former Duchy of Schleswig. Its capital city is Kiel; other notable cities are Lübeck, Flensburg and Neumünster.

The former English name was *Sleswick-Holsatia,* the Danish name is *Slesvig-Holsten,* the Low German name is *Sleswig-Holsteen,* the Dutch name is *Sleeswijk-Holstein* and the North Frisian name is *Slaswik-Holstiinj.* Historically, the name can also refer to a larger region, containing both present-day Schleswig-Holstein and the former South Jutland County (Northern Schleswig) in Denmark.

History

The term "Holstein" derives from Old Saxon, *Holseta Land,* meaning *"the land of those who dwell in the wood"* (*Holz* and Holt mean wood in modern Standardised German and in literary English respectively). Originally, it referred to the central of the three Saxon tribes north of the Elbe river, Tedmarsgoi, Holcetae, and Sturmarii. The area of the *Holcetae* was between the Stör river and Hamburg, and after Christianization their main church was in Schenefeld. Saxon Holstein became a part of the Holy Roman Empire after Charlemagne's Saxon campaigns in the late eighth century. Since 811 the northern frontier of Holstein (and thus the Empire) was marked by the river Eider.

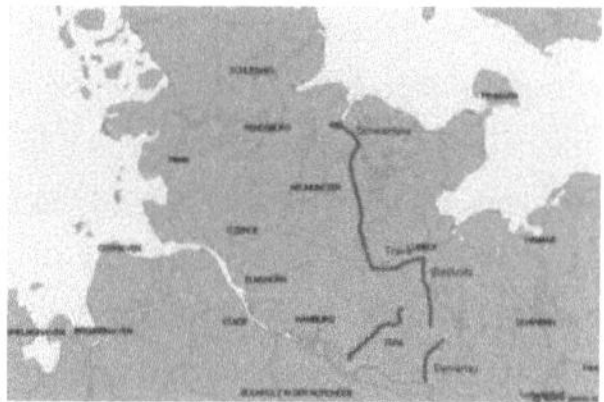

The *Limes Saxoniae* border between the Saxons and the Obotrites, established about 810 in present-day Schleswig-Holstein

The term *Schleswig* takes its name from the city of Schleswig. The name derives from the Schlei inlet in the east and *vik* meaning inlet or settlement in Old Saxon and Old Norse. The name is similar to the place-names ending in the "-wick" or "-wich" element along the coast in the United Kingdom.

Kiel is the state's capital and largest city.

The Duchy of Schleswig or Southern Jutland was originally an integral part of Denmark, but was in medieval times established as a fief under the Kingdom of Denmark, with the same relation to the Danish Crown as for example Brandenburg or Bavaria vis-à-vis the Holy Roman Emperor. Around 1100 the Duke of Saxony gave Holstein, as it was his own country, to Count Adolf I of Schauenburg.

Schleswig and Holstein have at different times belonged in part or completely to either Denmark or Germany, or have been virtually independent of both nations. The exception is that Schleswig had never been part of Germany until the Second War of Schleswig in 1864. For many centuries, the King of Denmark was both a Danish Duke of Schleswig and a German Duke of Holstein, the Duke of Saxony. Essentially, Schleswig was either integrated into Denmark or was a Danish fief, and Holstein was a German fief and once a sovereign state long ago. Both were for several centuries ruled by the kings of Denmark. In 1721 all of Schleswig was united as a single duchy under the king of Denmark, and the great powers of Europe confirmed in an international treaty that all future kings of Denmark should automatically become dukes of Schleswig, and consequently Schleswig would always follow the same line of succession as the one chosen in the Kingdom of Denmark.

The City of Lübeck was the centre of the Hanse, and its city centre is a World Heritage Site today. Lübeck is the birthplace of the author Thomas Mann.

The German national awakening following the Napoleonic Wars led to a strong popular movement in Holstein and Southern Schleswig for unification with a new Prussian-dominated Germany. However, this development was paralleled by an equally strong Danish national awakening in Denmark and northern Schleswig. It called for the complete reintegration of Schleswig into the Kingdom of Denmark and demanded an end to discrimination against Danes in Schleswig. The ensuing conflict is sometimes called the Schleswig-Holstein Question. In 1848 King Frederick VII of Denmark declared that he would grant Denmark a liberal constitution and the immediate goal for the Danish national movement was to ensure that this constitution would not only give rights to all Danes, i.e., not only in the Kingdom of Denmark, but also to Danes (and Germans) living in Schleswig. Furthermore, they demanded protection for the Danish language in Schleswig since the dominant language in almost a quarter of Schleswig had changed from Danish to German since the beginning of the 19th century.

A rapeseed field in Schleswig-Holstein — agriculture continues to play an important role in parts of the state.

A liberal constitution for Holstein was not seriously considered in Copenhagen, since it was a well-known fact that the political élite of Holstein had been far more conservative than Copenhagen's. This proved to be true, as the politicians of Holstein demanded that the Constitution of Denmark be scrapped — not only in Schleswig but also in Denmark. They also demanded that Schleswig immediately follow Holstein and become a member of the German Confederation, and eventually a part of the new united Germany. These demands were rejected and in 1848 the Germans of Holstein and Southern Schleswig rebelled. This was the beginning of the First War of Schleswig (1848–51) which ended in a Danish victory at Idstedt.

Schleswig-Holstein's islands, beaches and cities are popular tourist attractions (here: Isle of Sylt).

In 1863 conflict broke out again as King Frederick VII of Denmark died leaving no heir. According to the line of succession of Denmark and Schleswig, the crowns of both Denmark and Schleswig would now pass to Duke Christian of Glücksburg (the future King Christian IX); the crown of Holstein was considered to be more problematic. This decision was challenged by a rival pro-German branch of the Danish royal family, the House of Augustenborg who demanded, as in 1848, the crowns of both Schleswig and Holstein. The passing of a common constitution for Denmark and Schleswig in November 1863 then gave Otto von Bismarck a chance to intervene and Prussia and Austria declared war on Denmark. This was the Second War of Schleswig which ended in a Danish defeat. British attempts to mediate failed, and Denmark lost Schleswig (Northern and Southern Schleswig), Holstein, and Lauenburg to Prussia and Austria.

Following the Austro-Prussian War of 1866, section five of the Peace of Prague stated that the people in northern Schleswig should be granted the right to a referendum on whether they would remain under Prussian rule or return to Danish rule. This promise was never fulfilled by Prussia.

Following the defeat of Germany in World War I, the Allied powers arranged a plebiscite in northern and central Schleswig. In northern Schleswig (10 February 1920) 75% voted for reunification with Denmark and 25% voted for Germany. In central Schleswig (14 March 1920) the results were reversed; 80% voted for Germany and just 20% for Denmark, primarily in Flensburg. No vote ever took place in the southern third of Schleswig, although it was planned. For the referendum under authority of an international commission (CIS, Commission Internationale de Surveillance du Plébiscite au Slesvig) two (primarily three) election-zones were created. Primarily three zones were planned, Zone III should involve the rest of Southern Schleswig. Denmark passed on an election in this zone. Just the votes for the whole zone were crucial, not the dissenting votes in a single Kreis (district) or city:[5]

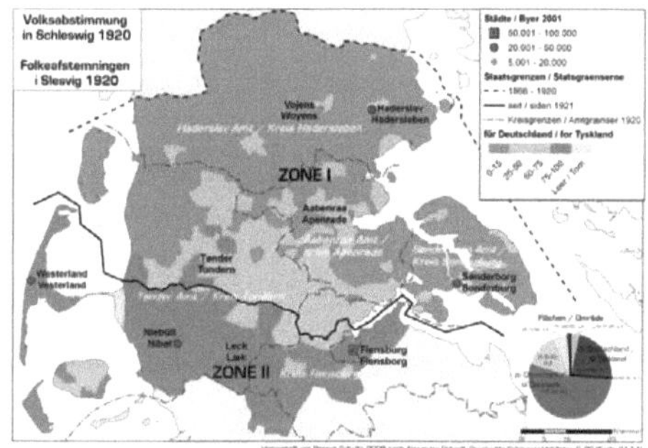

Results of the 1920 plebiscites in North and Central Schleswig (Slesvig)

Electorate	German name	Danish name	For Germany		For Denmark	
			percent	votes	percent	votes
Zone I (Northern Schleswig), 10 February 1920			**25.1 %**	**25,329**	**74.9 %**	**75,431**
District of	Hadersleben	Haderslev	16.0 %	6,585	84.0 %	34,653
Town of	Hadersleben	Haderslev	38.6 %	3,275	61.4 %	5,209
District of	Apenrade	Aabenraa	32.3 %	6,030	67.7 %	12,653
Town of	Apenrade	Aabenraa	55.1 %	2,725	44.9 %	2,224
District of	Sonderburg	Sønderborg	22.9 %	5,083	77.1 %	17,100
Town of	Sonderburg	Sønderborg	56.2 %	2,601	43.8 %	2,029
Town of	Augustenburg	Augustenborg	48.0 %	236	52.0 %	256
Northern part of District of	Tondern	Tønder	40.9 %	7,083	59.1 %	10,223
Town of	Tondern	Tønder	76.5 %	2,448	23.5 %	750
Town of	Hoyer	Højer	72.6 %	581	27.4 %	219
Town of	Lügumkloster	Løgumkloster	48.8 %	516	51.2 %	542
Northern part of District of	Flensburg	Flensborg	40.6 %	548	59.4 %	802
Zone II (Central Schleswig), 14 March 1920			**80.2 %**	**51,742**	**19.8 %**	**12,800**

Southern part of District of	Tondern	Tønder	87.9 %	17,283	12.1 %	2,376
Southern part of District of	Flensburg	Flensborg	82.6 %	6,688	17.4 %	1,405
Town of	Flensburg	Flensborg	75.2 %	27,081	24.8 %	8,944
Northern part of District of	Husum	Husum	90.0 %	672	10.0 %	75

On 15 June 1920, northern Schleswig officially returned to Danish rule. The Danish/German border was the only one of the borders imposed on Germany by the Treaty of Versailles after World War I which was never challenged by Adolf Hitler.

In 1937 the Nazis passed the so-called Greater Hamburg Act (*Groß-Hamburg-Gesetz*), where the nearby Free and Hanseatic City of Hamburg was expanded, to encompass towns that had formally belonged to the Prussian province of Schleswig-Holstein. To compensate Prussia for these losses (and partly because Hitler had a personal dislike for Lübeck), the 711-year-long independence of the Hansestadt Lübeck came to an end, and almost all its territory was incorporated into Schleswig-Holstein.

After the Second World War, the Prussian province Schleswig-Holstein came under British occupation. On August 23, 1946, the Military Government abolished the province and reconstituted it as a separate *Land*.[6]

Because of the Expulsion of Germans after World War II the population of Schleswig-Holstein increased 33 percent (860,000 people).[7]

Geography

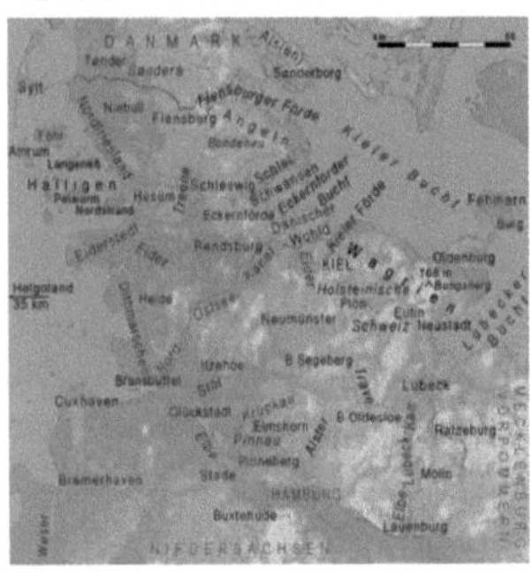

Geography

Schleswig-Holstein lies on the base of Jutland Peninsula between the North Sea and the Baltic Sea. Strictly speaking, "Schleswig" refers to the German Southern Schleswig, whereas Northern Schleswig is in Denmark. The state of Schleswig-Holstein further consists of Holstein as well as Lauenburg, and the formerly independent city of Lübeck.

Schleswig-Holstein borders Denmark (Region Syddanmark) to the north, the North Sea to the west, the Baltic Sea to the east, and the German states of Lower Saxony, Hamburg, and Mecklenburg-Vorpommern to the south.

In the western part of the state, there are lowlands with virtually no hills. The North Frisian Islands, as well as almost all of Schleswig-Holstein's North Sea coast, form the Schleswig-Holstein Wadden Sea National Park (*Nationalpark Schleswig-Holsteinisches Wattenmeer*) which is the largest national park in Central Europe. Germany's only high-sea island, Heligoland, is situated in the North Sea.

The Baltic Sea coast in the east of Schleswig-Holstein is marked by bays, fjords and cliff lines. There are rolling hills (the highest elevation is the Bungsberg at 168 metres or **unknown operator: u'strong'** feet) and many lakes, especially in the eastern part of Holstein, called the *Holsteinische Schweiz* ("Holsatian Switzerland") and the former Duchy of Lauenburg (*Herzogtum Lauenburg*). Fehmarn is the only island off the eastern coast. The longest river besides the Elbe is the Eider; the most important waterway is the Kiel Canal which connects the North Sea and Baltic Sea.

Administration

Schleswig-Holstein is divided into 11 *Kreise* (districts):

Districts

- Dithmarschen
- Lauenburg (formally *Herzogtum Lauenburg* or "Duchy of Lauenburg")
- Nordfriesland
- Ostholstein
- Pinneberg
- Plön
- Rendsburg-Eckernförde
- Schleswig-Flensburg
- Segeberg
- Steinburg
- Stormarn

Furthermore, there are four separate urban districts:

1. KI - Kiel
2. HL - *Hansestadt* ("Hanseatic town") Lübeck
3. NMS - Neumünster
4. FL - Flensburg

Demographics

Religion

The region has been strongly Protestant since the time of the Reformation. Today members of the Evangelical Church in Germany make up 54.3% of the population,[8] while members of the Catholic Church comprise 6%.[9]

Culture

Schleswig-Holstein combines Danish and German aspects of culture. The castles and manors in the countryside are the best example for this tradition; some dishes like Rote Grütze are also shared.

The most important festivals are the Schleswig-Holstein Musik Festival, an annual classic music festival all over the state, and the Nordische Filmtage, an annual film festival for movies from Scandinavian countries, held in Lübeck.

The annual Wacken Open Air festival is considered to be the largest heavy metal rock festival in the world.

Shared with the Danish neighbour: Rote Grütze served in Schleswig-Holstein with milk or custard

The state's most important museum of cultural history is in Schloss Gottorf in Schleswig.

Symbols

The coat of arms shows the symbols of the two duchies united in Schleswig-Holstein, i.e., the two lions for Schleswig and the leaf of a nettle for Holstein. Supposedly, Otto von Bismarck decreed that the two lions were to face the nettle because of the discomfort to their bottoms which would have resulted if the lions faced away from it.

The motto of Schleswig-Holstein is *"Up ewich ungedeelt"* (Middle Low German: "Forever undivided", modern High German: *"Auf ewig ungeteilt"*). It goes back to the Vertrag von Ripen or Handfeste von Ripen (Danish: Ribe Håndfæstning) or Treaty of Ribe in 1460. Ripen (Ribe) is a historical small town at the North Sea coast in Northern Schleswig. See History of Schleswig-Holstein.

The anthem from 1844 is called "Wanke nicht, mein Vaterland" ("Don't falter, my fatherland"), but it is usually referred to with its first line *"Schleswig-Holstein meerumschlungen"* (i.e., "Schleswig-Holstein embraced by the seas") or "Schleswig-Holstein-Lied" (Schleswig-Holstein song).

The old city of Lübeck is a world heritage site.

Languages

German is the official language, Low German, Danish and North Frisian enjoy legal protection or state promotion.

Historically, Low German, Danish (in Schleswig) and Frisian (in Schleswig) were spoken. Low German is still used in many parts of the state, and a pidgin of Low and standardised German (Missingsch) is used in most areas. Danish is used by the Danes in Southern Schleswig, and Frisian is spoken by the North Frisians of the North Sea Coast and the Northern Frisian Islands in Southern Schleswig. The North Frisian dialect called Heligolandic (*Halunder*) is spoken on the island of Heligoland.

Helgoland island in the North Sea

High German was introduced in the 16th century, mainly for official purposes, but is today the predominant language.

Education

Compulsory education starts for children who are six years old on June 30.[10] All children attend a "Grundschule", which is Germany's equivalent to primary school, for the first 4 years and then move on to a secondary school.[10] In Schleswig-Holstein there are "Gemeinschaftsschulen", which is a new type comprehensive school, as well as regional schools, which go by the German name "Regionalschule".[10] The option of a Gymnasium is still available.[10]

There are 3 universities in Kiel, Lübeck and Flensburg.[11] Also, there are 4 public Universities of Applied Sciences in Flensburg, Heide, Kiel, and Lübeck.[11] There is the Conservatory in Lübeck and the Muthesius Academy of Fine Arts in Kiel. There are also 3 private institutions of high learning.[11]

Politics

Schleswig-Holstein has its own parliament and government which are located in the state capital Kiel.[12] The Minister-President of Schleswig-Holstein is elected by the Landtag of Schleswig-Holstein.[12]

Current executive branch

Position	Minister	Party	Source
Minister-President	Peter Harry Carstensen	CDU	[13]
Minister of Agriculture, the Environment and Rural Areas	Dr. Juliane Rumpf	CDU	[14]
Minister of Education and Culture	Dr. Ekkehard Klug	FDP	[15]
Minister of Employment, Social Affairs and Health	Dr. Heiner Garg	FDP	[16]
Minister of Finance	Rainer Wiegard	CDU	[17]
Minister of the Interior	Klaus Schlie	CDU	[18]
Minister of Justice, Equality and Integration	Emil Schmalfuß	Ind	[19]
Minister of Science, Economic Affairs and Transport	Jost de Jager	CDU	[20]

Last election

The last Schleswig-Holstein state election was held on 27 September 2009, and the result of it was a coalition of the conservative CDU and the liberal FDP under the leadership of CDU state premier Peter Harry Carstensen. It was an early election; after the Christian Democratic Union (CDU) and Social Democratic Party (SPD) grand coalition broke apart in Summer 2009, Minister-President Peter Harry Carstensen (CDU) provoked early elections by intentionally losing a vote of confidence.

Political Party	Votes %	+/-	Seats
Christian Democratic Union (*Christlich Demokratische Union Deutschlands*)	31.5	-8.7	34
Social Democratic Party of Germany (*Sozialdemokratische Partei Deutschlands*)	25.4	-12.3	25
Free Democratic Party of Germany (*Freie Demokratische Partei*)	14.9	+8.3	14
Alliance '90/The Greens (*Bündnis 90/Die Grünen*)	12.4	+2.2	12
The Left (*Die Linke*)	6.0	+5.2	6
South Schleswig Voter Federation (*Südschleswigscher Wählerverband*)	4.3	+0.7	4
Pirate Party Germany (*Piratenpartei Deutschland*)	1.8	+1.8	-
Free Voters (*Freie Wähler*)	1.0	+1.0	-
National Democratic Party of Germany (*Nationaldemokratische Partei Deutschlands*)	0.9	-1.0	-

See also

- Outline of Germany
- Schleswig-Holstein Question
- First Schleswig War
- Second Schleswig War
- Schleswig
- Holstein
- Duchy of Schleswig
- Holstein-Glückstadt
- Dukes of Holstein-Gottorp
- Schleswig-Holstein-Sonderburg
- Schleswig-Holstein-Sonderburg-Glücksburg
- Schleswig-Holstein-Sonderburg-Beck
- Schleswig-Holstein-Sonderburg-Augustenburg
- Schleswig-Holstein-Sonderburg-Plön
- Schleswig-Holstein-Sonderburg-Norburg
- Schleswig-Holstein-Sonderburg-Plön-Rethwisch
- Coat of arms of Schleswig
- Region Sønderjylland-Schleswig

References

[1] "State population" (http://www.statistik-portal.de/Statistik-Portal/de_zs01_shs.asp). *Portal of the Federal Statistics Office Germany.* . Retrieved 2007-04-25.

[2] By the federal vehicle registration reform of 1 July 1956 distinct prefixes were given for every district.

[3] "Schleswig-Holstein: Schleswig-Holstein Portal" (http://www.schleswig-holstein.de/). Schleswig-holstein.de. . Retrieved 2010-04-14.

[4] http://www.schleswig-holstein.de/

[5] Schwedler, Frank: Historischer Atlas Schleswig-Holstein 1867 bis 1945, Wachholtz Verlag, Neumünster

[6] Ordinance No. 46, *Abolition of the Provinces in the British Zone of the Former State of Prussia and Reconstitution thereof as Separate Länder* (http://www.lwl.org/westfaelische-geschichte/que/normal/que1167.pdf)PDF (218 KB)

[7] Flucht und Vertreibung (http://www.hdg.de/lemo/html/Nachkriegsjahre/DasEndeAlsAnfang/fluchtUndVertreibung.html) at Haus der Geschichte (German)

[8] EKD http://www.ekd.de/download/kirchenmitglieder_2007.pdf

[9] chiesa cattolica http://www.dbk.de/imperia/md/content/kirchlichestatistik/bev-kath-l__nd-2008.pdf

[10] "Education in Schleswig-Holstein" (http://www.schleswig-holstein.de/Portal/EN/Education/Education_node.html). State of Schleswig-Holstein. . Retrieved 14 April 2011.

[11] "Institutions of Higher Education in Schleswig-Holstein" (http://www.schleswig-holstein.de/Portal/EN/Education/InstitutionsHigherEducation/InstitutionsHigherEducation_node.html). State of Schleswig-Holstein. . Retrieved 14 April 2011.
[12] "Responsibilities of the Government" (http://www.schleswig-holstein.de/Portal/EN/LandGovernment/TheResponsibilities/Responsibilities_node.html). State of Schleswig-Holstein. . Retrieved 14 April 2011.
[13] "Ministerpräsident Peter Harry Carstensen im Porträt" (http://www.schleswig-holstein.de/STK/DE/Ministerpraesident/Portraet/Portraet_node.html) (in German). State of Schleswig-Holstein. . Retrieved 14 April 2011.
[14] "Ministry of Agriculture, the Environment and Rural Areas" (http://www.schleswig-holstein.de/Portal/EN/LandGovernment/StateChancelleryMinistries/MinistryAgricultureEnvironmentRuralAreas/MinistryAgricultureEnvironmentRuralAreas_node.html). State of Schleswig-Holstein. . Retrieved 14 April 2011.
[15] "Ministry of Education and Culture" (http://www.schleswig-holstein.de/Portal/EN/LandGovernment/StateChancelleryMinistries/MinistryEducationCulture/MinistryEducationCulture_node.html). State of Schleswig-Holstein. . Retrieved 14 April 2011.
[16] "Ministry of Employment, Social Affairs and Health" (http://www.schleswig-holstein.de/Portal/EN/LandGovernment/StateChancelleryMinistries/MinistryEmploymentSocialAffairsHealth/MinistryEmploymentSocialAffairsHealth_node.html). State of Schleswig-Holstein. . Retrieved 14 April 2011.
[17] "Ministry of Finance" (http://www.schleswig-holstein.de/Portal/EN/LandGovernment/StateChancelleryMinistries/MinistryFinance/MinistryFinance_node.html). State of Schleswig-Holstein. . Retrieved 14 April 2011.
[18] "Ministry of the Interior" (http://www.schleswig-holstein.de/Portal/EN/LandGovernment/StateChancelleryMinistries/MinistryInterior/MinistryInterior_node.html). State of Schleswig-Holstein. . Retrieved 14 April 2011.
[19] "Ministry of Justice, Equality and Integration" (http://www.schleswig-holstein.de/Portal/EN/LandGovernment/StateChancelleryMinistries/MinistryJusticeEqualityIntegration/MinistryJusticeEqualityIntegration_node.html). State of Schleswig Holstein. . Retrieved 14 April 2011.
[20] "Ministry of Science, Economic Affairs and Transport" (http://www.schleswig-holstein.de/Portal/EN/LandGovernment/StateChancelleryMinistries/MinistryScienceEconomicAffairsTransport/MinistryScienceEconomicAffairsTransport_node.html). State of Schleswig-Holstein. . Retrieved 14 April 2011.

External links

- Official government portal (http://www.schleswig-holstein.de/)
- Official Directory (http://www.sh-regional.de/)
- Schleswig-Holstein Plebiscite Paper Money (http://www.numismondo.com/pm/sch/) - 1919, 1920 Issues
- 360° Panoramas of Schleswig-Holstein (http://www.schleswig-holstein-360.de)
- Chisholm, Hugh, ed. (1911). "Schleswig-Holstein". *Encyclopædia Britannica* (11th ed.). Cambridge University Press.

Germany

Federal Republic of Germany *Bundesrepublik Deutschland*	

Anthem: The third stanza of *Das Lied der Deutschen* *The Song of the Germans*	
 Location of Germany(dark green) – in Europe(green & dark grey) – in the European Union(green) — [Legend]	
Capital (and largest city)	Berlin 52°31′N 13°23′E
Official language(s)	German[1]
Demonym	German
Government	Federal parliamentary constitutional republic
- President	Joachim Gauck
- Chancellor	Angela Merkel
- President of the Bundestag	Norbert Lammert
- President of the Bundesrat	Horst Seehofer
Formation	
- Holy Roman Empire	2 February 962
- Unification	18 January 1871
- Federal Republic	23 May 1949

-	Reunification	3 October 1990
Area		
-	Total	357,021 km² (63rd) 137,847 sq mi
-	Water (%)	2.416
Population		
-	2010 estimate	81,799,600[1] (16th)
-	Density	229/km² (55th) 593/sq mi
GDP (PPP)		2011 estimate
-	Total	$3.089 trillion[2] (5th)
-	Per capita	$37,935[2] (18th)
GDP (nominal)		2011 estimate
-	Total	$3.628 trillion[2] (4th)
-	Per capita	$44,555[2] (19th)
Gini (2006)		27 (low)
HDI (2011)		▲ 0.905[3] (very high) (9th)
Currency		Euro (€)[2](2002 – present) (EUR)
Time zone		CET (UTC+1)
-	Summer (DST)	CEST (UTC+2)
Drives on the		right
ISO 3166 code		DE
Internet TLD		.de [3]
Calling code		49
1	Danish, Low German, Sorbian, Romany and Frisian are officially recognised by the ECRML.	
2	Before 2002: Deutsche Mark (DEM).	
3	Also .eu, shared with European Union member states.	

Germany (i/ˈdʒɜrməni/), officially the **Federal Republic of Germany** (German: *Bundesrepublik Deutschland*, pronounced [ˈbʊndəsʁepuˌbliːk ˈdɔʏtʃlant] (listen)),[4] is a federal parliamentary republic in Europe. The country consists of 16 states while the capital and largest city is Berlin. Germany covers an area of 357,021 km² and has a largely temperate seasonal climate. With 81.8 million inhabitants, it is the most populous member state and the largest economy in the European Union. It is one of the major political powers of the European continent and a technological leader in many fields.

A region named Germania, inhabited by several Germanic peoples, was documented before AD 100. During the Migration Period, the Germanic tribes expanded southward, and established successor kingdoms throughout much of Europe. Beginning in the 10th century, German territories formed a central part of the Holy Roman Empire.[5] During the 16th century, northern German regions became the centre of the Protestant Reformation while southern and western parts remained dominated by Roman Catholic denominations, with the two factions clashing in the

Thirty Years' War, marking the beginning of the Catholic–Protestant divide that has characterized German society ever since.[6] Occupied during the Napoleonic Wars, the rise of Pan-Germanism inside the German Confederation resulted in the unification of most of the German states into the German Empire in 1871 which was Prussian dominated. After the German Revolution of 1918–1919 and the subsequent military surrender in World War I, the Empire was replaced by the Weimar Republic in 1918, and partitioned in the Treaty of Versailles. Amidst the Great Depression, the Third Reich was proclaimed in 1933. The latter period was marked by Fascism and World War II. After 1945, Germany was divided by allied occupation, and evolved into two states, East Germany and West Germany. In 1990 Germany was reunified.

Germany was a founding member of the European Community in 1957, which became the EU in 1993. It is part of the Schengen Area and since 1999 a member of the eurozone. Germany is a member of the United Nations, NATO, the G8, the G20, the OECD and the Council of Europe, and took a non-permanent seat on the UN Security Council for the 2011–2012 term.

It has the world's fourth largest economy by nominal GDP and the fifth largest by purchasing power parity. It is the second largest exporter and third largest importer of goods. The country has developed a very high standard of living and a comprehensive system of social security. Germany has been the home of many influential scientists and inventors, and is known for its cultural and political history.

Etymology

The English word *Germany* derives from the Latin Germania, which came into use after Julius Caesar adopted it for the peoples east of the Rhine.[7] In other languages it has various names.

The German term *Deutschland* (originally *diutisciu land*, "the German lands") is derived from *deutsch*, descended from Old High German *diutisc* "popular" (i. e., belonging to the *diot* or *diota* "people"; originally used to distinguish the language of the common people from Latin and its Romance descendants). This in turn descends from Proto-Germanic **þiudiskaz* "popular" (see also the Latinised form Theodiscus), derived from **þeudō*, descended from Proto-Indo-European *$tewt\acute{e}h_2$- "people".[8]

History

Germanic tribes and Frankish Empire

Map of the Germania and the Roman Empire

The Germanic tribes are thought to date from the Nordic Bronze Age or the Pre-Roman Iron Age. From southern Scandinavia and north Germany, they expanded south, east and west from the 1st century BC, coming into contact with the Celtic tribes of Gaul as well as Iranian, Baltic, and Slavic tribes in Eastern Europe.[9] Under Augustus, the Roman General Publius Quinctilius Varus began to invade Germania (an area extending roughly from the Rhine to the Ural Mountains). In AD 9, three Roman legions led by Varus were defeated by the Cheruscan leader Arminius. By AD 100, when Tacitus wrote *Germania*, Germanic tribes had settled along the Rhine and the Danube (the Limes Germanicus), occupying most of the area of modern Germany; Austria, southern Bavaria and the western Rhineland, however, were Roman provinces.[10]

In the 3rd century a number of large West Germanic tribes emerged: Alemanni, Franks, Chatti, Saxons, Frisii, Sicambri, and Thuringii. Around 260, the Germanic peoples broke into Roman-controlled lands.[11] After the invasion of the Huns in 375, and with the decline of Rome from 395, Germanic tribes moved further south-west. Simultaneously several large tribes formed in what is now Germany and displaced the smaller Germanic tribes. Large areas (known since the Merovingian period as Austrasia) were occupied by the Franks, and Northern Germany was ruled by the Saxons and Slavs.[10]

Holy Roman Empire

Martin Luther initiated the Protestant Reformation.

On 25 December 800, Charlemagne founded the Carolingian Empire, which was divided in 843.[12] The Holy Roman Empire resulted from the eastern portion of this division. Its territory stretched from the Eider River in the north to the Mediterranean coast in the south.[12] Under the reign of the Ottonian emperors (919–1024), several major duchies were consolidated, and the German king was crowned Holy Roman Emperor of these regions in 962. The Holy Roman Empire absorbed northern Italy and Burgundy under the reign of the Salian emperors (1024–1125), although the emperors lost power through the Investiture Controversy.

Under the Hohenstaufen emperors (1138–1254), the German princes increased their influence further south and east into territories inhabited by Slavs, preceding German settlement in these areas and further east *(Ostsiedlung)*. Northern German towns grew prosperous as members of the Hanseatic League.[13] Starting with the Great Famine in 1315, then the Black Death of 1348–50, the population of Germany plummeted.[14] The edict of the Golden Bull in 1356 provided the basic constitution of the empire and codified the election of the emperor by seven prince-electors who ruled some of the most powerful principalities and archbishoprics.[15]

Martin Luther publicised his 95 Theses in 1517, challenging the Roman Catholic Church and initiating the Protestant Reformation. A separate Lutheran church became the official religion in many German states after 1530. Religious conflict led to the Thirty Years' War (1618–1648), which devastated German lands.[16] The population of the German states was reduced by about 30%.[17] The Peace of Westphalia (1648) ended religious warfare among the German states, but the empire was *de facto* divided into numerous independent principalities. From 1740 onwards, dualism between the Austrian Habsburg Monarchy and the Kingdom of Prussia dominated German history. In 1806, the *Imperium* was overrun and dissolved as a result of the Napoleonic Wars.[18]

German Confederation and Empire

Following the fall of Napoleon I of France, the Congress of Vienna convened in 1814 and founded the German Confederation (Deutscher Bund), a loose league of 39 sovereign states. Disagreement with restoration politics partly led to the rise of liberal movements, followed by new measures of repression by Austrian statesman Metternich. The *Zollverein*, a tariff union, furthered economic unity in the German states.[19] National and liberal ideals of the French Revolution gained increasing support among many, especially young, Germans. In the light of a series of revolutionary movements in Europe, which established a republic in France, intellectuals and commoners started the Revolutions of 1848 in the German states. King Frederick William IV of Prussia was offered the title of Emperor, but with a loss of power; he rejected the crown and the proposed constitution, leading to a temporary setback for the movement.[20]

Foundation of the German Empire in Versailles, 1871. Bismarck is at the centre in a white uniform.

Conflict between King William I of Prussia and the increasingly liberal parliament erupted over military reforms in 1862, and the king appointed Otto von Bismarck the new Prime Minister of Prussia. Bismarck successfully waged war on Denmark in 1864. Prussian victory in the Austro-Prussian War of 1866 enabled him to create the North German Federation (Norddeutscher Bund) and to exclude Austria, formerly the leading German state, from the federation's affairs. After the French defeat in the Franco-Prussian War, the German Empire was proclaimed 1871 in Versailles, uniting all scattered parts of Germany except Austria (*Kleindeutschland*, or "Lesser Germany"). With almost two thirds of its territory and population, Prussia was the dominating constituent of the new state; the Hohenzollern King of Prussia ruled as its concurrent Emperor, and Berlin became its capital.[20] In the *Gründerzeit* period following the unification of Germany, Bismarck's foreign policy as Chancellor of Germany under Emperor William I secured Germany's position as a great nation by forging alliances, isolating France by diplomatic means, and avoiding war. Under Wilhelm II, however, Germany, like other European powers, took an imperialistic course leading to friction with neighbouring countries. As a result of the Berlin Conference in 1884 Germany claimed several colonies including German East Africa, German South-West Africa, Togo, and Cameroon.[21] Most alliances in which Germany had previously been involved were not renewed, and new alliances excluded the country.[22]

The assassination of Austria's crown prince on 28 June 1914 triggered World War I. Germany, as part of the Central Powers, suffered defeat against the Allies in one of the bloodiest conflicts of all time. An estimated two million German soldiers died in World War I.[23] The German Revolution broke out in November 1918, and Emperor Wilhelm II and all German ruling princes abdicated. An armistice ended the war on 11 November, and Germany was forced to sign the Treaty of Versailles in June 1919. The treaty was perceived in Germany as a humiliating continuation of the war, and is often cited as an influence in the rise of Nazism.[24]

Weimar Republic and Nazi Regime

Adolf Hitler, chancellor and president[1] 1933–19451: office formally vacant from August 1934; Hitler styled himself *"Führer und Reichskanzler"*.[25]

At the beginning of the German Revolution in November 1918, Germany was declared a republic. However, the struggle for power continued, with radical-left communists seizing power in Bavaria. The revolution came to an end on 11 August 1919, when the democratic Weimar Constitution was signed by President Friedrich Ebert.[26] Suffering from the Great Depression of 1929, the harsh peace conditions dictated by the Treaty of Versailles, and a long succession of unstable governments, Germans increasingly lacked identification with the government. This was exacerbated by a widespread right-wing *Dolchstoßlegende*, or stab-in-the-back myth, which argued that Germany had lost World War I because of those who wanted to overthrow the government. The Weimar government was accused of betraying Germany by signing the Versailles Treaty. By 1932, the German Communist Party and the Nazi Party controlled the majority of parliament, fuelled by discontent with the Weimar government. After a series of unsuccessful cabinets, President Paul von Hindenburg appointed Adolf Hitler as Chancellor of Germany on 30 January 1933.[27] On 27 February 1933 the Reichstag building went up in flames, and a consequent emergency decree abrogated basic citizens' rights. An Enabling Act passed in parliament gave Hitler unrestricted legislative power. Only the Social Democratic Party voted against it, while Communist MPs had already been imprisoned.[28] [29] Using his powers to crush any actual or potential resistance, Hitler established a centralised totalitarian state within months. Industry was revitalised with a focus on military rearmament.[30]

In 1935, Germany reacquired control of the Saar and in 1936 military control of the Rhineland, both of which had been lost in the Treaty of Versailles.[31] In 1938, Austria was annexed, and in 1939, Czechoslovakia was brought under German control. The invasion of Poland was prepared through the Molotov–Ribbentrop pact and Operation Himmler. On 1 September 1939 the German Wehrmacht launched a blitzkrieg on Poland, which was swiftly occupied by Germany and by the Soviet Red Army. The UK and France declared war on Germany, marking the beginning of World War II.[32] As the war progressed, Germany and its allies quickly gained control of most of continental Europe and North Africa, though plans to force the United Kingdom to an armistice or surrender failed. On 22 June 1941, Germany broke the Molotov–Ribbentrop pact and invaded the Soviet Union. Japan's attack on Pearl Harbor led Germany to declare war on the United States. The Battle of Stalingrad forced the German army to retreat on the Eastern front.[32] In September 1943, Germany's ally Italy surrendered, and German troops were forced to defend an additional front in Italy. D-Day opened a Western front, as Allied forces advanced towards German territory. On 8 May 1945, the German armed forces surrendered after the Red Army occupied Berlin.[33]

Berlin in ruins after World War II

In what later became known as The Holocaust, the Third Reich regime had enacted policies directly subjugating many dissidents and minorities. Millions of people were murdered by the Nazis during the Holocaust, including approximately six million Jews, and a sizeable number of Romani people, Jehovah's Witnesses, Poles and other Slavs, Soviet POWs, people with mental and/or physical disabilities, homosexuals, and members of the political and religious opposition.[34] World War II was responsible for more than 40 million dead in Europe.[35] The Nuremberg trials of Nazi war criminals were held after World War II.[36] The war casualties for Germany are estimated at 5.3

million German soldiers[37] millions of German civilians;[38] [39] [40] [41] [42] and losing the war resulted in large territorial losses; the expulsion of about 15 million Germans from the eastern areas of Germany and other countries; mass rape of German women;[43] and the destruction of multiple major cities.

East and West Germany

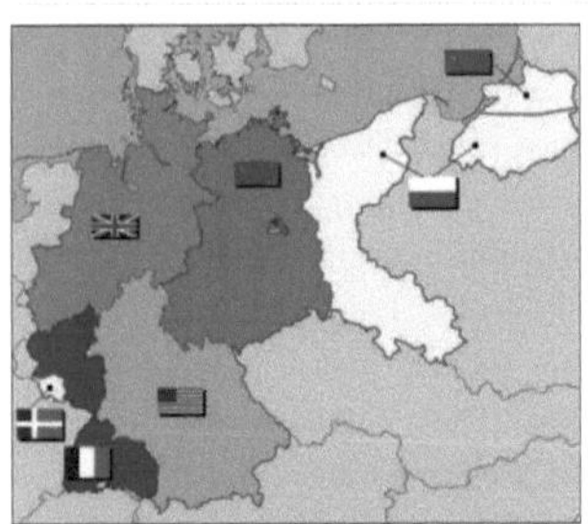

Occupation zones in Germany, 1947. The territories east of the Oder-Neisse line, under Polish and Soviet *de jure* administration and *de facto* annexation, are shown as white as is the detached Saar protectorate.

After the surrender of Germany, the remaining German territory and Berlin were partitioned by the Allies into four military occupation zones. The western sectors, controlled by France, the United Kingdom, and the United States, were merged on 23 May 1949 to form the *Federal Republic of Germany* (*Bundesrepublik Deutschland*); on 7 October 1949, the Soviet Zone became the *German Democratic Republic* (*Deutsche Demokratische Republik*, or DDR). They were informally known as "West Germany" and "East Germany". East Germany selected East Berlin as its capital, while West Germany chose Bonn as a provisional capital, to emphasise its stance that the two-state solution was an artificial and temporary *status quo*.[44]

West Germany, established as a federal parliamentary republic with a "social market economy", was allied with the United States, the UK and France. The country enjoyed prolonged economic growth beginning in the early 1950s (*Wirtschaftswunder*). West Germany joined NATO in 1955 and was a founding member of the European Economic Community in 1957. East Germany was an Eastern bloc state under political and military control by the USSR via the latter's occupation forces and the Warsaw Pact. Though East Germany claimed to be a democracy, political power was exercised solely by leading members (*Politbüro*) of the communist-controlled Socialist Unity Party of Germany (SED), supported by the Stasi, an immense secret service,[45] and a variety of sub-organisations controlling every aspect of society. A Soviet-style command economy was set up; the GDR later became a Comecon state.[46] While East German propaganda was based on the benefits of the GDR's social programmes and the alleged constant threat of a West German invasion, many of her citizens looked to the West for freedom and prosperity.[47] The Berlin Wall, built in 1961 to stop East Germans from escaping to West Germany, became a symbol of the Cold War.[20]

The Berlin Wall in front of the Brandenburg Gate shortly before its fall in 1989

Tensions between East and West Germany were reduced in the early 1970s by Chancellor Willy Brandt's *Ostpolitik*. In summer 1989, Hungary decided to dismantle the Iron Curtain and open the borders, causing the emigration of thousands of East Germans to West Germany via Hungary. This had devastating effects on the GDR, where regular mass demonstrations received increasing support. The East German authorities unexpectedly eased the border restrictions, allowing East German citizens to travel to the West; originally intended to help retain East Germany as a state, the opening of the border actually led to an acceleration of the *Wende* reform process. This culminated in the *Two Plus Four Treaty* a year later on 12 September 1990, under which the four occupying powers renounced their rights under the Instrument of Surrender, and Germany regained full sovereignty. This permitted German reunification on 3 October 1990, with the accession of the five re-established states of the former GDR (new states or "neue Länder").[20]

Berlin Republic and the EU

Based on the Berlin/Bonn Act, adopted on 10 March 1994, Berlin once again became the capital of the reunified Germany, while Bonn obtained the unique status of a *Bundesstadt* (federal city) retaining some federal ministries.[48] The relocation of the government was completed in 1999.[49] Since reunification, Germany has taken a more active role in the European Union and NATO. Germany sent a peacekeeping force to secure stability in the Balkans and sent a force of German troops to Afghanistan as part of a NATO effort to provide security in that country after the ousting of the Taliban.[50] These deployments were controversial since, after the war, Germany was bound by domestic law only to deploy troops for defence roles.[51] In 2005, Angela Merkel became the first female Chancellor of Germany as the leader of a grand coalition.[20]

Geography

Germany is in Western and Central Europe, bordering Denmark in the north, Poland and the Czech Republic in the east, Austria and Switzerland in the south, France and Luxembourg in the south-west, and Belgium and the Netherlands in the north-west. It lies mostly between latitudes 47° and 55° N (the tip of Sylt is just north of 55°), and longitudes 5° and 16° E. The territory covers 357021 km^2 (**unknown operator: u'strong'** sq mi), consisting of 349223 km^2 (**unknown operator: u'strong'** sq mi) of land and 7798 km^2 (**unknown operator: u'strong'** sq mi) of water. It is the seventh largest country by area in Europe and the 62nd largest in the world.[52]

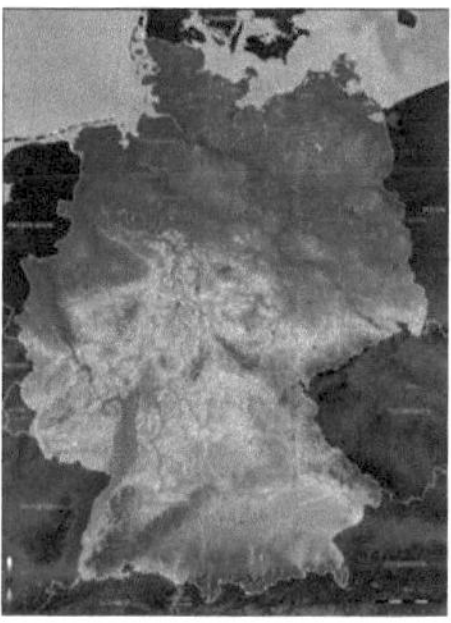

Topographic map

Elevation ranges from the mountains of the Alps (highest point: the Zugspitze at 2962 metres / **unknown operator: u'strong'** feet) in the south to the shores of the North Sea (Nordsee) in the north-west and the Baltic Sea (Ostsee) in the north-east. The forested uplands of central Germany and the lowlands of northern Germany (lowest point: Wilstermarsch at 3.54 metres / **unknown operator: u'strong'** feet below sea level) are traversed by such major rivers as the Rhine, Danube and Elbe. Glaciers are found in the Alpine region, but are experiencing deglaciation. Significant natural resources are iron ore, coal, potash, timber, lignite, uranium, copper, natural gas, salt, nickel, arable land and water.[52]

Climate

Most of Germany has a temperate seasonal climate in which humid westerly winds predominate. The climate is moderated by the North Atlantic Drift, the northern extension of the Gulf Stream. This warmer water affects the areas bordering the North Sea; consequently in the north-west and the north the climate is oceanic. Rainfall occurs year-round, especially in the summer. Winters are mild and summers tend to be cool, though temperatures can exceed 30 °C (**unknown operator: u'strong'** °F).[53]

The east has a more continental climate; winters can be very cold and summers very warm, and long dry periods are frequent. Central and southern Germany are transition regions which vary from moderately oceanic to continental. In addition to the maritime and continental climates that predominate over most of the country, the Alpine regions in the extreme south and, to a lesser degree, some areas of the Central German Uplands have a mountain climate, characterised by lower temperatures and greater precipitation.[53]

Biodiversity

The eagle is a protected bird of prey and the national heraldic animal.

The territory of Germany can be subdivided into two ecoregions: European-Mediterranean montane mixed forests and Northeast-Atlantic shelf marine.[54] As of 2008 the majority of Germany is covered by either arable land (34%) or forest and woodland (30.1%); only 13.4% of the area consists of permanent pastures, 11.8% is covered by settlements and streets.[55]

Plants and animals are those generally common to middle Europe. Beeches, oaks, and other deciduous trees constitute one third of the forests; conifers are increasing as a result of reforestation. Spruce and fir trees predominate in the upper mountains, while pine and larch are found in sandy soil. There are many species of ferns, flowers, fungi, and mosses. Wild animals include deer, wild boar, mouflon, fox, badger, hare, and small numbers of beavers.[56]

The national parks in Germany include the Wadden Sea National Parks, the Jasmund National Park, the Vorpommern Lagoon Area National Park, the Müritz National Park, the Lower Oder Valley National Park, the Harz National Park, the Saxon Switzerland National Park and the Bavarian Forest National Park. More than 400 registered zoos and animal parks operate in Germany, which is believed to be the largest number in any country.[57] The Zoologische Garten Berlin is the oldest zoo in Germany and presents the most comprehensive collection of species in the world.[58]

Politics

The Reichstag in Berlin is the site of the German parliament (Bundestag).

Germany is a federal, parliamentary, representative democratic republic. The German political system operates under a framework laid out in the 1949 constitutional document known as the *Grundgesetz* (Basic Law). Amendments generally require a two-thirds majority of both chambers of parliament; the fundamental principles of the constitution, as expressed in the articles guaranteeing human dignity, the separation of powers, the federal structure, and the rule of law are valid in perpetuity.[59]

The president is the head of state and invested primarily with representative responsibilities and powers. He is elected by the *Bundesversammlung* (federal convention), an institution consisting of the members of the *Bundestag* and an equal number of state delegates. The second-highest official in the German order of precedence is the *Bundestagspräsident* (President of the *Bundestag*), who is elected by the *Bundestag* and responsible for overseeing the daily sessions of the body. The third-highest official and the head of government is the Chancellor, who is appointed by the *Bundespräsident* after being elected by the *Bundestag*.[20]

The chancellor, currently Angela Merkel, is the head of government and exercises executive power, similar to the role of a Prime Minister in other parliamentary democracies. Federal legislative power is vested in the parliament consisting of the *Bundestag* (Federal Diet) and *Bundesrat* (Federal Council), which together form the legislative body. The *Bundestag* is elected through direct elections, by proportional representation (mixed-member).[52] The members of the *Bundesrat* represent the governments of the sixteen federated states and are members of the state cabinets.[20]

Joachim Gauck, President of Germany

Since 1949, the party system has been dominated by the Christian Democratic Union and the Social Democratic Party of Germany with all chancellors hitherto being member of either party. However, the smaller liberal Free Democratic Party (which has had members in the *Bundestag* since 1949) and the Alliance '90/The Greens (which has controlled seats in parliament since 1983) have also played important roles.[60]

Germany has a civil law system based on Roman law with some references to Germanic law. The *Bundesverfassungsgericht* (Federal Constitutional Court) is the German Supreme Court responsible for constitutional matters, with power of judicial review.[20] [61] Germany's supreme court system, called *Oberste Gerichtshöfe des Bundes*, is specialised: for civil and criminal cases, the highest court of appeal is the inquisitorial Federal Court of Justice, and for other affairs the courts are the Federal Labour Court, the Federal Social Court, the Federal Finance Court and the Federal Administrative Court. The *Völkerstrafgesetzbuch* regulates the consequences of crimes against humanity, genocide and war crimes, and gives German courts universal jurisdiction in some circumstances.[62] Criminal and private laws are codified on the national level in the *Strafgesetzbuch* and the *Bürgerliches Gesetzbuch* respectively. The German penal system is aimed towards rehabilitation of the criminal and the protection of the general public.[63] Except for petty crimes, which are tried before a single professional judge, and serious political crimes, all charges are tried before mixed tribunals on which lay judges (*Schöffen*) sit side by side with professional judges.[64] [65]

Constituent states

Germany comprises sixteen states that are collectively referred to as *Länder*.[66] Each state has its own state constitution[67] and is largely autonomous in regard to its internal organisation. Due to differences in size and population the subdivision of these states varies, especially between city states (*Stadtstaaten*) and states with larger territories (*Flächenländer*). For regional administrative purposes five states, namely Baden-Württemberg, Bavaria, Hesse, North Rhine-Westphalia and Saxony, consist of a total of 22 Government Districts (*Regierungsbezirke*). As of 2009 Germany is divided into 403 districts (*Kreise*) on municipal level, these consist of 301 rural districts and 102 urban districts.[68]

Lower Saxony

Bremen

Hamburg

Mecklenburg-Vorpommern

Saxony-Anhalt

Saxony

Brandenburg

Berlin

Thuringia

Hesse

North Rhine-Westphalia

Rhineland-Palatinate

Bavaria

Baden-Württemberg

Saarland

Schleswig-Holstein

State	Capital	Area (km²)	Population
Baden-Württemberg	Stuttgart	35,752	10,717,000
Bavaria	Munich	70,549	12,444,000
Berlin	Berlin	892	3,400,000
Brandenburg	Potsdam	29,477	2,568,000
Bremen	Bremen	404	663,000
Hamburg	Hamburg	755	1,735,000
Hesse	Wiesbaden	21,115	6,098,000
Mecklenburg-Vorpommern	Schwerin	23,174	1,720,000
Lower Saxony	Hanover	47,618	8,001,000
North Rhine-Westphalia	Düsseldorf	34,043	18,075,000
Rhineland-Palatinate	Mainz	19,847	4,061,000
Saarland	Saarbrücken	2,569	1,056,000
Saxony	Dresden	18,416	4,296,000
Saxony-Anhalt	Magdeburg	20,445	2,494,000
Schleswig-Holstein	Kiel	15,763	2,829,000
Thuringia	Erfurt	16,172	2,355,000

Foreign relations

Germany has a network of 229 diplomatic missions abroad and maintains relations with more than 190 countries.[69] As of 2011 it is the largest contributor to the budget of the European Union (providing 20%)[70] and the third largest contributor to the UN (providing 8%).[71] Germany is a member of NATO, the Organisation of Economic Co-operation and Development (OECD), the G8, the G20, the World Bank and the International Monetary Fund (IMF). It has played a leading role in the European Union since its inception and has maintained a strong alliance with France since the end of World War II. Germany seeks to advance the creation of a more unified European political, defence, and security apparatus.[72] [73]

Chancellor Angela Merkel hosting the G8 summit in Heiligendamm

The development policy of the Federal Republic of Germany is an independent area of German foreign policy. It is formulated by the Federal Ministry for Economic Cooperation and Development (BMZ) and carried out by the implementing organisations. The German government sees development policy as a joint responsibility of the international community.[74] It is the world's third biggest aid donor after the United States and France.[75] [76]

During the Cold War, Germany's partition by the Iron Curtain made it a symbol of East-West tensions and a political battleground in Europe. However, Willy Brandt's Ostpolitik was a key factor in the *détente* of the 1970s.[77] In 1999, Chancellor Gerhard Schröder's government defined a new basis for German foreign policy by taking part in the NATO decisions surrounding the Kosovo War and by sending German troops into combat for the first time since World War II.[78] The governments of Germany and the United States are close political allies.[20] The 1948 Marshall Plan and strong cultural ties have crafted a strong bond between the two countries, although Schröder's vocal opposition to the Iraq War suggested the end of Atlanticism and a relative cooling of German-American relations.[79] The two countries are also economically interdependent: 8.8% of German exports are U.S.-bound and

6.6% of German imports originate from the U.S.[80]

Military

Germany's military, the *Bundeswehr*, is organized in *Heer* (Army), *Marine* (Navy), *Luftwaffe* (Air Force), *Zentraler Sanitätsdienst* (Central Medical Services) and *Streitkräftebasis* (Joint Support Service) branches. As of 2005, military spending was an estimated 1.5% of the country's GDP, that is position 99 in a ranking of all countries;[52] absolutely, German military expenditure is the eighth-highest in the world.[81] In peacetime, the Bundeswehr is commanded by the Minister of Defence. If Germany went to war, which according to the constitution is allowed only for defensive purposes, the Chancellor would become commander in chief of the *Bundeswehr*.[82]

As of May 2011 the Bundeswehr employs 188,000 professional soldiers, 31,000 18–25 year-old conscripts who serve for at least six months.[83] The German government plans to reduce the number of soldiers to 170,000 professionals and up to 15,000 short-time volunteers (*voluntary military service*).[84] Reservists are available to the Armed Forces and participate in defence exercises and deployments abroad, a new reserve concept of their future strength and functions was announced 2011.[84] As of April 2011, the German military had about 6,900 troops stationed in foreign countries as part of international peacekeeping forces, including about 4,900 Bundeswehr troops in the NATO-led ISAF force in Afghanistan and Uzbekistan, 1,150 German soldiers in Kosovo, and 300 troops with UNIFIL in Lebanon.[85]

Until 2011, military service was compulsory for men at age 18, and conscripts served six-month tours of duty; conscientious objectors could instead opt for an equal length of *Zivildienst* (civilian service), or a six-year commitment to (voluntary) emergency services like a fire department or the Red Cross. On 1 July 2011 conscription was officially suspended and replaced with a voluntary service.[86] [87] Since 2001 women may serve in all functions of service without restriction, but they are not subject to conscription. There are presently some 17,500 women on active duty and a number of female reservists.[88]

Economy

Germany has a social market economy with a highly qualified labour force, a large capital stock, a low level of corruption,[90] and a high level of innovation.[91] It has the largest national economy in Europe, the fourth largest by nominal GDP in the world,[92] and the fifth largest by PPP[92] in 2009. The service sector contributes approximately 71% of the total GDP, industry 28%, and agriculture 0.9%.[52] The average national unemployment rate in 2010 was about 7.5%.[52]

A Mercedes-Benz car. Germany was the world's leading exporter of goods from 2003 to 2008.[89]

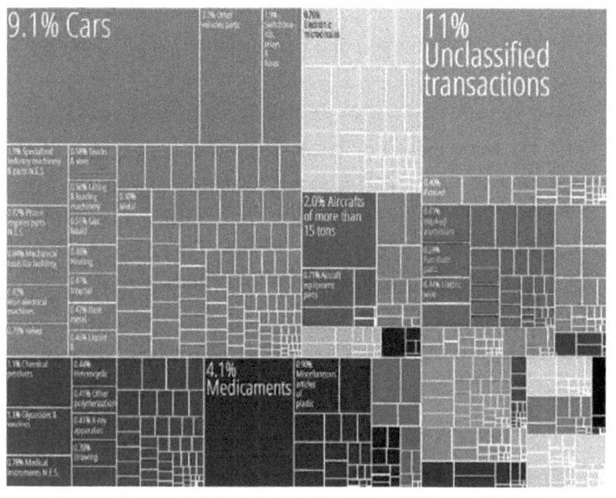

Germany Exports (2009) by Product Category. MIT/Harvard Economic Complexity Observatory.[93]

Germany is a founding member of the EU, the G8 and the G20, and was the world's largest exporter from 2003 to 2008. In 2009 it remained the second largest exporter and third largest importer of goods. Most of the country's exports are in engineering, especially machinery, automobiles, chemical goods and metals.[52] Germany is a leading producer of wind turbines and solar-power technology.[94] Annual trade fairs and congresses are held in cities throughout Germany.[95]

Germany is an advocate of closer European economic and political integration. Its commercial policies are increasingly determined by agreements among European Union (EU) members and by EU legislation. Germany introduced the common European currency, the euro, on 1 January 2002.[96] [97] Its monetary policy is set by the European Central Bank. Two decades after German reunification, standards of living and per capita incomes remain significantly higher in the states of the former West Germany than in the former East.[98] The modernisation and integration of the eastern German economy is a long-term process scheduled to last until the year 2019, with annual transfers from west to east amounting to roughly $80 billion.[99] In January 2009 the German government approved a €50 billion economic stimulus plan to protect several sectors from a downturn and a subsequent rise in unemployment rates.[100]

Of the world's 500 largest stock-market-listed companies measured by revenue in 2010, the Fortune Global 500, 37 are headquartered in Germany. 30 Germany-based companies are included in the DAX, the German stock market index. Well-known global brands are Mercedes-Benz, BMW, SAP, Siemens, Volkswagen, Adidas, Audi, Allianz, Porsche, and Nivea.[101] Germany is recognised for its specialised small and medium enterprises. Around 1,000 of these companies are global market leaders in their segment and are labelled hidden champions.[102]

The list includes the largest companies by turnover in 2009. Unranked are the largest bank and the largest insurance company in 2007:

Rank[103]	Name	Headquarters	Revenue (Mil. €)	Profit (Mil. €)	Employees (World)
1	Volkswagen AG	Wolfsburg	108,897	4,120	329,305
2	Daimler AG	Stuttgart	99,399	3,985	272,382
3	Siemens AG	Munich/Berlin	72,488	3,806	398,200
4	E.ON AG	Düsseldorf	68,731	7,204	87,815
5	Metro AG	Düsseldorf	64,337	825	242,378
6	Deutsche Post AG	Bonn	63,512	1,389	475,100
7	Deutsche Telekom AG	Bonn	62,516	569	241,426
8	BASF SE	Ludwigshafen	57,951	4,065	95,175
9	BMW AG	Munich	56,018	3,126	107,539
10	ThyssenKrupp AG	Essen/Duisburg	51,723	2,102	191,350

Infrastructure

The ICE 3 train

With its central position in Europe, Germany is a transport hub. This is reflected in its dense and modern transport networks. The motorway (Autobahn) network ranks as the third largest worldwide in length.[104] Germany has established a polycentric network of high-speed trains. The InterCityExpress or *ICE* network of the Deutsche Bahn serves major German cities as well as destinations in neighbouring countries with speeds up to 300 kph (186 mph).[105] The largest German airports are Frankfurt Airport and Munich Airport, both hubs of Lufthansa, while Air Berlin has hubs at Berlin Tegel and Düsseldorf. Other major airports include Berlin Schönefeld, Hamburg, Cologne/Bonn and Leipzig/Halle. Both airports in Berlin will be consolidated at a site adjacent to Berlin Schönefeld, which will become Berlin Brandenburg Airport in 2012.[106]

As of 2008, Germany was the world's sixth largest consumer of energy,[107] and 60% of its primary energy was imported.[108] Government policy promotes energy conservation and renewable energy. Energy efficiency has been improving since the early 1970s; the government aims to meet the country's electricity demands using only renewable sources by 2050.[109] In 2010, energy sources were: oil (33.7%); coal, including lignite (22.9%); natural gas (21.8%); nuclear (10.8%); hydro-electric and wind power (1.5%); and other renewable sources (7.9%).[110] In 2000, the government and the nuclear power industry agreed to phase out all nuclear power plants by 2021.[111] Germany is committed to the Kyoto protocol and several other treaties promoting biodiversity, low emission standards, recycling, and the use of renewable energy, and supports sustainable development at a global level.[112] The German government has initiated wide-ranging emission reduction activities and the country's overall emissions are falling.[113] Nevertheless the country's greenhouse gas emissions were the highest in the EU as of 2007.[114]

Science and technology

Albert Einstein

Germany's achievements in sciences have been significant, and research and development efforts form an integral part of the economy.[115] The Nobel Prize has been awarded to 103 German laureates.[116] For most of the 20th century, German laureates had more awards than those of any other nation, especially in the sciences (physics, chemistry, and physiology or medicine).[117] [118]

The work of Albert Einstein and Max Planck was crucial to the foundation of modern physics, which Werner Heisenberg and Max Born developed further.[119] Wilhelm Conrad Röntgen discovered X-rays and was the first winner of the Nobel Prize in Physics in 1901.[120] Numerous mathematicians were born in Germany, including Carl Friedrich Gauss, David Hilbert, Bernhard Riemann, Gottfried Leibniz, Karl Weierstrass, Hermann Weyl and Felix Klein. Research institutions in Germany include the Max Planck Society, the Helmholtz Association and the Fraunhofer Society. The Gottfried Wilhelm Leibniz Prize is granted to ten scientists and academics every year. With a maximum of €2.5 million per award it is one of highest endowed research prizes in the world.[121]

Germany has been the home of many famous inventors and engineers, such as Johannes Gutenberg, credited with the invention of movable type printing in Europe; Hans Geiger, the creator of the Geiger counter; and Konrad Zuse, who built the first fully automatic digital computer.[122] German inventors, engineers and industrialists such as Count Ferdinand von Zeppelin, Otto Lilienthal, Gottlieb Daimler, Rudolf Diesel, Hugo Junkers and Karl Benz helped shape modern automotive and air transportation technology.[123] Aerospace engineer Wernher von Braun developed the first space rocket and later on was a prominent member of NASA and developed the Saturn V Moon rocket, which paved the way for the success of the US Apollo program. Heinrich Rudolf Hertz's work in the domain of electromagnetic radiation was pivotal to the development of modern telecommunication.[124]

Germany is also one of the leading countries in developing and using green technologies. Companies specializing in green technology have an estimated turnover of 200€ billion. Especially the expertise in engineering, science and research of Germany is eminently respectable. The lead markets of Germany's green technology industry are power generation, sustainable mobility, material efficiency, energy efficiency, waste management and recycling, sustainable water management.[125]

Demographics

With its estimated population of 81.8 million in January 2010,[1] Germany is the most populous country in the European Union and ranks as the 16th most populous country in the world.[126] Its population density stands at 229.4 inhabitants per square kilometre. The overall life expectancy in Germany at birth is 79.9 years. The fertility rate of 1.4 children per mother, or 7.9 births per 1000 inhabitants in 2009, is one of the lowest in the world.[127] Since the 1970s, Germany's death rate has continuously exceeded its birth rate.[128] The Federal Statistical Office of Germany forecast that the population will shrink to between 65 and 70 million by 2060 (depending on the level of net migration).[129]

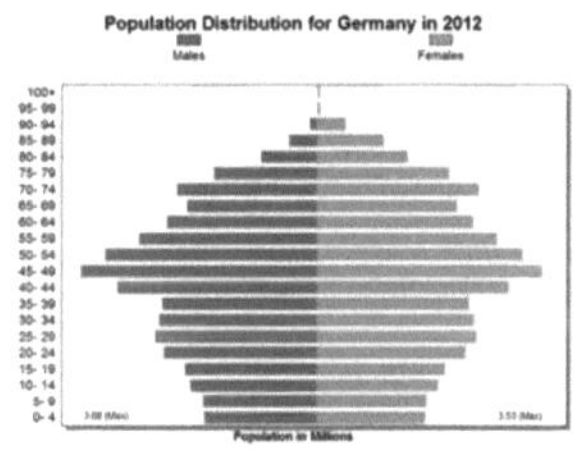

Population pyramid of Germany in 2012 from International Futures

German nationals make up 91% of the population of Germany. As of 2010, about seven million foreign citizens were registered in Germany, and 19% of the country's residents were of foreign or partially foreign descent (including persons descending or partially descending from ethnic German repatriates), 96% of whom lived in West Germany or Berlin.[130] The United Nations Population Fund lists Germany as host to the third-highest number of international migrants worldwide, about 5% or 10 million of all 191 million migrants.[131] As a consequence of restrictions to Germany's formerly rather unrestricted laws on asylum and immigration, the number of immigrants seeking asylum or claiming German ethnicity (mostly from the former Soviet Union) has been declining steadily since 2000.[132] In 2009, 20% of the population had immigrant roots, the highest since 1945.[133] As of 2008, the largest national group was from Turkey (2.5 million), followed by Italy (776,000) and Poland (687,000).[134] About 3 million "Aussiedler"—ethnic Germans, mainly from Eastern Europe and the former Soviet Union—have resettled in Germany since 1987.[135]

Germany has a number of large cities. The largest conurbation is the Rhine-Ruhr region (11.5 million as of 2006), including Düsseldorf (the capital of North Rhine-Westphalia), Cologne, Dortmund, Essen, Duisburg, and Bochum.[136]

Religion

Christianity is the largest religion in Germany, with around 51.5 million adherents (62.8%) in 2008,[137] of which 30.0% are Catholics and 29.9% are Protestants, belonging to the Evangelical Church in Germany (EKD); the remainder consists of small denominations (each less than 0.5% of the German population).[138] Protestantism is concentrated in the north and east and Roman Catholicism is concentrated in the south and west;[6] 1.6% of the country's overall population declare themselves Orthodox Christians.[137]

The Cologne Cathedral at the Rhine river is a UNESCO World Heritage Site.

The second largest religion is Islam with an estimated 3.8 to 4.3 million adherents (4.6% to 5.2%),[139] followed by Buddhism with 250,000 and Judaism with around 200,000 adherents (0.3%); Hinduism has some 90,000 adherents (0.1%). All other religious communities in Germany have fewer than 50,000 adherents.[140] Of the roughly 4 million Muslims, most are Sunnis and Alevites from Turkey, but there are a small number of Shi'ites and other denominations.[139] German Muslims, a large portion of whom are of Turkish origin, lack full official state recognition of their religious community.[6] Germany has Europe's third largest Jewish population (after France and the United Kingdom).[141] Approximately 50% of the Buddhists in Germany are Asian immigrants.[142]

Germans with no stated religious adherence make up 34.1% of the population, especially in the former East Germany and major metropolitan areas.[138] German reunification in 1990 greatly increased the country's non-religious population, a legacy of the state atheism of the previously Soviet-controlled East. Christian church membership has decreased in recent decades, particularly among Protestants.[6]

Languages

German is the official and predominant spoken language in Germany.[143] It is one of 23 official languages in the European Union, and one of the three working languages of the European Commission. Recognised native minority languages in Germany are Danish, Low German, Sorbian, Romany, and Frisian; they are officially protected by the ECRML. The most used immigrant languages are Turkish, Kurdish, Polish, the Balkan languages, and Russian; 67% of German citizens claim to be able to communicate in at least one foreign language and 27% in at least two languages other than their own.[143]

Standard German is a West Germanic language and is closely related to and classified alongside English, Low German, Dutch, and the Frisian languages. To a lesser extent, it is also related to the East (extinct) and North Germanic languages. Most German vocabulary is derived from the Germanic branch of the Indo-European language family.[144] Significant minorities of words are derived from Latin and Greek, with a smaller amount from French and most recently English (known as Denglisch). German is written using the Latin alphabet. German dialects, traditional local varieties traced back to the Germanic tribes, are distinguished from varieties of standard German by their lexicon, phonology, and syntax.[145]

Education

The University of Heidelberg was established in 1386.

Over 99% of Germans age 15 and above are estimated to be able to read and write.[52] However, a growing number of inhabitants are functionally illiterate.[146] Responsibility for educational oversight in Germany lies primarily with the individual federated states. Since the 1960s, a reform movement attempted to unify secondary education in a *Gesamtschule* (comprehensive school); several West German states later simplified their school system to two or three tiers. A system of apprenticeship called *Duale Ausbildung* ("dual education") allows pupils in vocational training to learn in a company as well as in a state-run vocational school.[147]

Optional kindergarten education is provided for all children between three and six years old, after which school attendance is compulsory for at least nine years. Primary education usually lasts for four years and public schools are not stratified at this stage.[147] In contrast, secondary education includes three traditional types of schools focused on different levels of academic ability: the *Gymnasium* enrols the most gifted children and prepares students for university studies; the *Realschule* for intermediate students lasts six years; the *Hauptschule* prepares pupils for vocational education.[148]

The general entrance requirement for university is Abitur, a qualification normally based on continuous assessment during the last few years at school and final examinations; however there are a number of exceptions, and precise requirements vary, depending on the state, the university and the subject. Germany's universities are recognised internationally; in the Academic Ranking of World Universities (ARWU) for 2008, six of the top 100 universities in the world are in Germany, and 18 of the top 200.[149] Nearly all German universities are public institutions, charging tuition fees of €50–500 per semester for each student.[150]

Health

Germany has the world's oldest universal health care system, dating back to Otto von Bismarck's Social legislation in 1883.[151] Currently the population is covered by a fairly comprehensive health insurance plan provided by statute. Certain groups of people (lifetime officials, self-employed persons, employees with high income) can opt out of the plan and switch to a private insurance contract. Previously, these groups could also choose to do without insurance, but this option was dropped in 2009.[152] According to the World Health Organization, Germany's health care system was 77% government-funded and 23% privately funded as of 2005.[153] In 2005, Germany spent 11% of its GDP on health care. Germany ranked 20th in the world in life expectancy with 77 years for men and 82 years for women, and it had a very low infant mortality rate (4 per 1,000 live births).[153]

As of 2010, the principal cause of death was cardiovascular disease, at 41.1%, followed by malignant tumours, at 25.5%.[154] As of 2008, about 82,000 Germans had been infected with HIV/AIDS and 26,000 had died from the disease (cumulatively, since 1982).[155] According to a 2005 survey, 27% of German adults are smokers.[155]

Culture

Ludwig van Beethoven (1770–1827), composer

From its roots, culture in Germany has been shaped by major intellectual and popular currents in Europe, both religious and secular. Historically Germany has been called *Das Land der Dichter und Denker* (the land of poets and thinkers).[156] The federated states are in charge of the cultural institutions. There are 240 subsidised theatres, hundreds of symphonic orchestras, thousands of museums and over 25,000 libraries spread in Germany. These cultural opportunities are enjoyed by many: there are over 91 million German museum visits every year; annually, 20 million go to theatres and operas; 3.6 million per year listen to the symphonic orchestras.[157] The UNESCO inscribed 33 properties in Germany on the World Heritage List.[158]

Germany has established a high level of gender equality,[159] promotes disability rights, and is legally and socially tolerant towards homosexuals. Gays and lesbians can legally adopt their partner's biological children, and civil unions have been permitted since 2001.[160] Germany has also changed its attitude towards immigrants; since the mid-1990s, the government and the majority of Germans have begun to acknowledge that controlled immigration should be allowed based on qualification standards.[161] Germany has been named the world's second most valued nation among 50 countries in 2010.[162] A global opinion poll for the BBC revealed that Germany is recognised for having the most positive influence in the world in 2011.[163]

Arts

J.S.Bach	L.v. Beethoven	R. Wagner
Toccata und Fuge	Symphonie 5 c-moll	Die Walküre

Numerous German painters have enjoyed international prestige through their work in diverse artistic styles. Hans Holbein the Younger, Matthias Grünewald, and Albrecht Dürer were important artists of the Renaissance, Caspar David Friedrich of Romanticism, and Max Ernst of Surrealism. Architectural contributions from Germany include the Carolingian and Ottonian styles, which were precursors of Romanesque. The region later became the site of Gothic, Renaissance and Baroque art. Germany was particularly important in the early modern movement, especially through the Bauhaus movement founded by Walter Gropius. Ludwig Mies van der Rohe became one of the world's most renowned architects in the second half of the 20th century. He conceived of the glass façade skyscraper.[164]

German music includes works by some of the world's most well-known classical music composers, including Ludwig van Beethoven, Johann Sebastian Bach, Wolfgang Amadeus Mozart, Johannes Brahms, and Richard Wagner. As of 2008, Germany is the fourth largest music market in the world[165] and has influenced popular music through artists such as Kraftwerk, Alphaville, Boney M., Nena, Nico, Nina Hagen, Scorpions, Die Toten Hosen, Tokio Hotel, Rammstein, and Paul van Dyk.

Literature and philosophy

The Brothers Grimm

German literature can be traced back to the Middle Ages and the works of writers such as Walther von der Vogelweide and Wolfram von Eschenbach. Well-known German authors include Johann Wolfgang von Goethe and Friedrich Schiller. The collections of folk tales published by the Brothers Grimm popularised German folklore on an international level. Influential authors of the 20th century include Thomas Mann, Bertolt Brecht, Hermann Hesse, Heinrich Böll, and Günter Grass.[166] German-speaking book publishers produce some 700 million books every year, with about 80,000 titles, nearly 60,000 of them new. Germany comes third in quantity of books published, after the English-speaking book market and the People's Republic of China.[167] The Frankfurt Book Fair is the most important in the world for international deals and trading, with a tradition spanning over 500 years.[168]

German philosophy is historically significant. Gottfried Leibniz's contributions to rationalism; the establishment of classical German idealism by Immanuel Kant, Johann Gottlieb Fichte, Georg Wilhelm Friedrich Hegel and Friedrich Wilhelm Joseph Schelling; Arthur Schopenhauer's composition of metaphysical pessimism; the formulation of communist theory by Karl Marx and Friedrich Engels; Friedrich Nietzsche's development of perspectivism; Gottlob Frege's contributions to the dawn of analytic philosophy; Martin Heidegger's works on Being; and the development of the Frankfurt school by Max Horkheimer, Theodor Adorno, Herbert Marcuse and Jürgen Habermas have been particularly influential. In the 21st century Germany has contributed to the development of contemporary analytic philosophy in continental Europe, along with France, Austria, Switzerland and the Scandinavian countries.[169]

Media

German cinema dates back to the earliest years of the medium with the work of Max Skladanowsky, which was particularly influential with German expressionists such as Robert Wiene and Friedrich Wilhelm Murnau. Director Fritz Lang's *Metropolis* (1927) is referred to as the first modern science-fiction film. In 1930 the Austrian-American Josef von Sternberg directed *The Blue Angel*, the first major German sound film.[170] During the 1970s and 1980s, New German Cinema directors such as Volker Schlöndorff, Werner Herzog, Wim Wenders, and Rainer Werner Fassbinder put West German cinema on the international stage.[171] The annual European Film Awards ceremony is held every other year in Berlin, home of the European Film Academy (EFA); the Berlin Film Festival, held annually since 1951, is one of the world's foremost film festivals.[172]

More recently, films such as *Good Bye Lenin!* (2003), *Gegen die Wand (Head-on)* (2004), *Der Untergang (Downfall)* (2004), and *Der Baader Meinhof Komplex* (2008) have had international success. The Academy Award for Best Foreign Language Film went to the German production *Die Blechtrommel (The Tin Drum)* in 1979, to *Nowhere in Africa* in 2002, and to *Das Leben der Anderen (The Lives of Others)* in 2007.[173] Germany's television market is the largest in Europe, with some 34 million TV households. Around 90% of German households have cable or satellite TV, with a variety of free-to-view public and commercial channels.[174]

Cuisine

German cuisine varies from region to region. The southern regions of Bavaria and Swabia, for instance, share a culinary culture with Switzerland and Austria. In all regions, meat is often eaten in sausage form.[175] Organic food has gained a market share of ca. 2%, and is expected to increase further.[176] Although wine is becoming more popular in many parts of Germany, the national alcoholic drink is beer. German beer consumption per person is declining, but at 121.4 litres in 2009 it is still among the highest in the world.[177] The Michelin guide has awarded nine restaurants in Germany three stars, the highest designation, while 15 more received two stars.[178] German restaurants have become the world's second-most decorated after France.[179]

A *Schwarzwälder Kirschtorte* (literally, "Black Forest cherry torte".)

Sports

Signal Iduna Park is the biggest stadium in Germany

Twenty-seven million Germans are members of a sports club and an additional twelve million pursue sports individually.[180] Association football is the most popular sport. With more than 6.3 million official members, the German Football Association (*Deutscher Fußball-Bund*) is the largest sports organisation of its kind worldwide.[180] The Bundesliga attracts the second highest average attendance of any professional sports league in the world.

The German national football team won the FIFA World Cup in 1954, 1974 and 1990 and the UEFA European Football Championship in 1972, 1980 and 1996. Germany hosted the FIFA World Cup in 1974 and 2006 and the UEFA European Football Championship in 1988. Among the most well-known footballers are Franz Beckenbauer, Gerd Müller, Jürgen Klinsmann, Lothar Matthäus, and Oliver Kahn. Other popular spectator sports include handball, volleyball, basketball, ice hockey, and tennis.[180]

Germany is one of the leading motor sports countries in the world. Constructors like BMW and Mercedes are prominent manufacturers in motor sport. Additionally, Porsche has won the 24 Hours of Le Mans, an annual endurance race held in France, 16 times, and Audi has won it 9 times. Formula One driver Michael Schumacher has set many motor sport records during his career, having won more Formula One World Drivers' Championships and more Formula One races than any other driver; he is one of the highest paid sportsmen in history.[181]

Historically, German sportsmen have been successful contenders in the Olympic Games, ranking third in an all-time Olympic Games medal count, combining East and West German medals. In the 2008 Summer Olympics, Germany finished fifth in the medal count,[182] while in the 2006 Winter Olympics they finished first.[183] Germany has hosted the Summer Olympic Games twice, in Berlin in 1936 and in Munich in 1972. The Winter Olympic Games took place in Germany once in 1936 in the twin towns of Garmisch and Partenkirchen.

References

[1] *Key Figures on Europe* (http://epp.eurostat.ec.europa.eu/cache/ITY_OFFPUB/KS-EI-11-001/EN/KS-EI-11-001-EN.PDF). Belgium: European Union. 2011. p. 37. doi:10.2785/623. ISBN 978-92-79-18441-3. .

[2] "World Economic Outlook Database" (http://www.imf.org/external/pubs/ft/weo/2011/02/weodata/weorept.aspx?pr.x=41&pr.y=5&sy=2009&ey=2016&scsm=1&ssd=1&sort=country&ds=.&br=1&c=134&s=NGDPD,NGDPDPC,PPPGDP,PPPPC,LP&grp=0&a=). International Monetary Fund. September 2011. . Retrieved 17-November-2011.

[3] "Human development index" (http://hdr.undp.org/en/media/HDR_2011_EN_Table1.pdf). United Nations Development Programme. 2011. . Retrieved 5 November 2011.

[4] Mangold, Max, ed. (1995) (in German). *Duden, Aussprachewörterbuch* (6th ed.). Dudenverlag. pp. 271, 53f. ISBN 9783411209163.

[5] The Latin name *Sacrum Imperium* (Holy Empire) is documented as far back as 1157. The Latin name *Sacrum Romanum Imperium* (Holy Roman Empire) was first documented in 1254. The full name "Holy Roman Empire of the German Nation" (*Heiliges Römisches Reich Deutscher Nation*) dates back to the 15th century.
Zippelius, Reinhold (2006) [1994] (in German). *Kleine deutsche Verfassungsgeschichte: vom frühen Mittelalter bis zur Gegenwart [Brief German Constitutional History: from the Early Middle Ages to the Present]* (7 ed.). Munich: Beck. p. 25. ISBN 9783406476389.

[6] "Germany" (http://berkleycenter.georgetown.edu/resources/countries/germany). Berkley Center for Religion, Peace, and World Affairs. . Retrieved 2011-12-15.

[7] Schulze, Hagen (1998). *Germany: A New History*. Harvard University Press. p. 4. ISBN 0-674-80688-3.

[8] Lloyd, Albert L.; Lühr, Rosemarie; Springer, Otto (1998) (in German). *Etymologisches Wörterbuch des Althochdeutschen, Band II* (http://books.google.com/books?id=iKfYGNwwNVIC&pg=PA523). Göttingen: Vandenhoeck & Ruprecht. pp. 699–704. ISBN 3-525-20768-9. . (for *diutisc*) Lloyd, Albert L.; Lühr, Rosemarie; Springer, Otto (1998) (in German). *Etymologisches Wörterbuch des Althochdeutschen, Band II* (http://books.google.com/books?id=iKfYGNwwNVIC&pg=PA516). Göttingen: Vandenhoeck & Ruprecht. pp. 685–686. ISBN 3-525-20768-9. . (for *diot*)

[9] Claster, Jill N. (1982). *Medieval Experience: 300–1400*. New York University Press. p. 35. ISBN 0-8147-1381-5.

[10] Fulbrook 1991, pp. 9–13.

[11] Bowman, Alan K.; Garnsey, Peter; Cameron, Averil (2005). *The crisis of empire, A.D. 193–337*. The Cambridge Ancient History. **12**. Cambridge University Press. p. 442. ISBN 0-521-30199-8.

[12] Fulbrook 1991, p. 11.

[13] Fulbrook 1991, pp. 13–24.

[14] Nelson, Lynn Harry. *The Great Famine (1315–1317) and the Black Death (1346–1351)* (http://www.vlib.us/medieval/lectures/black_death.html). University of Kansas. . Retrieved 19 March 2011.

[15] Fulbrook 1991, p. 27.

[16] Philpott, Daniel (January 2000). "The Religious Roots of Modern International Relations". *World Politics* **52** (2): 206–245.

[17] Macfarlane, Alan (1997). *The savage wars of peace: England, Japan and the Malthusian trap*. Blackwell. p. 51. ISBN 9780631181170.

[18] Fulbrook 1991, p. 97.

[19] Henderson, W. O. (January 1934). "The Zollverein". *History* **19** (73): 1–19. doi:10.1111/j.1468-229X.1934.tb01791.x.

[20] "Germany" (http://www.state.gov/r/pa/ei/bgn/3997.htm). U.S. Department of State. 10 November 2010. . Retrieved 26 March 2011.

[21] Black, John, ed. (2005). *100 maps*. Sterling Publishing. p. 202. ISBN 9781402728853.

[22] Fulbrook 1991, pp. 135, 149.

[23] Crossland, David (22 January 2008). "Last German World War I Veteran Believed to Have Died" (http://www.spiegel.de/international/germany/0,1518,530319,00.html). *Spiegel Online*. . Retrieved 25 March 2011.

[24] Lee, Stephen J. (2003). *Europe, 1890–1945*. Routledge. p. 131. ISBN 9780415254557.

[25] Thamer, Hans-Ulrich (2003). "Beginn der nationalsozialistischen Herrschaft (Teil 2)" (http://www.bpb.de/geschichte/nationalsozialismus/dossier-nationalsozialismus/39548/gleichschaltung) (in German). *Nationalsozialismus I*. Bonn: Federal Agency for Civic Education. . Retrieved 5 April 2012. "President von Hindenburg died on August 2nd, 1934. The day before, the cabinet had approved a submission making Hitler his successor. The office of the president was to be dissolved and united with that of the chancellor under the name "Führer und Reichskanzler". However, this was in breach of the Enabling Act. (shortened & paraphrased)"

[26] Fulbrook 1991, pp. 156–160.

[27] Fulbrook 1991, pp. 155–158, 172–177.

[28] "Das Ermächtigungsgesetz 1933" (http://www.dhm.de/lemo/html/nazi/innenpolitik/ermaechtigungsgesetz/index.html) (in German). Deutsches Historisches Museum. . Retrieved 25 March 2011.

[29] Stackelberg, Roderick (1999). *Hitler's Germany: Origins, interpretations, legacies*. Routledge. p. 103. ISBN 9780415201155.

[30] "Industrie und Wirtschaft" (http://www.dhm.de/lemo/html/nazi/wirtschaft/index.html) (in German). Deutsches Historisches Museum. . Retrieved 25 March 2011.

[31] Fulbrook 1991, pp. 188–189.

[32] Fulbrook 1991, pp. 190–195.

[33] Steinberg, Heinz Günter (1991) (in German). *Die Bevölkerungsentwicklung in Deutschland im Zweiten Weltkrieg: mit einem Überblick über die Entwicklung von 1945 bis 1990*. Kulturstiftung der dt. Vertriebenen. ISBN 9783885570899.

[34] Niewyk, Donald L.; Nicosia, Francis R. (2000). *The Columbia Guide to the Holocaust*. Columbia University Press. pp. 45–52. ISBN 9780231112000.

[35] "Leaders mourn Soviet wartime dead" (http://news.bbc.co.uk/2/hi/europe/4530565.stm). *BBC News*. 9 May 2005. . Retrieved 18 March 2011.

[36] Overy, Richard (17 February 2011). "Nuremberg: Nazis on Trial" (http://www.bbc.co.uk/history/worldwars/wwtwo/nuremberg_article_01.shtml). BBC History. . Retrieved 25 March 2011.

[37] Rűdiger Overmans. *Deutsche militärische Verluste im Zweiten Weltkrieg*. Oldenbourg 2000. ISBN 3-486-56531-1

[38] Das Deutsche Reich und der Zweite Weltkrieg, Bd. 9/1, ISBN 3-421-06236-6. Page 460 (This study was prepared by the German Armed Forces Military History Research Office, an agency of the German government)

[39] Peter Antill; Peter Dennis (10 October 2005). *Berlin 1945: end of the Thousand Year Reich* (http://books.google.com/books?id=vAzgsCDUky0C). Osprey Publishing. p. 85. ISBN 9781841769158. . Retrieved 24 June 2011.

[40] Bonn : Kulturstiftung der Deutschen Vertriebenen, *Vertreibung und Vertreibungsverbrechen, 1945–1948 : Bericht des Bundesarchivs vom 28. Mai 1974 : Archivalien und ausgewählte Erlebnisberichte* / [Redaktion, Silke Spieler]. Bonn :1989 ISBN 388557067X. (This is a study of German expulsion casualties due to "war crimes" prepared by the German government Archives)

[41] Germany reports. With an introd. by Konrad Adenauer. Germany (West). Presse- und Informationsamt. Wiesbaden, Distribution: F. Steiner, 1961] Page 32

[42] Robert N. Proctor, *Racial Hygiene: Medicine under the Nazis*, Harvard 1988,

[43] Beevor, Antony (2003) [2002]. *Berlin: The downfall 1945*. Penguin. pp. 31–32, 409–412. ISBN 9780140286960.

[44] Wise, Michael Z. (1998). *Capital dilemma: Germany's search for a new architecture of democracy*. Princeton Architectural Press. p. 23. ISBN 978-1-56898-134-5.

[45] maw/dpa (11 March 2008). "New Study Finds More Stasi Spooks" (http://www.spiegel.de/international/germany/0,1518,540771,00.html). *Spiegel Online – english site (www.spiegel.de/international)*. Der Spiegel. . Retrieved 30 October 2011. "189,000 people were informers for the Stasi – the former Communist secret police – when East Germany collapsed in 1989 – 15,000 more than previous studies had suggested. [...] about one in 20 members of the former East German Communist party, the SED, was a secret police informant."

[46] Colchester, Nico (1 January 2001). "D-mark day dawns" (http://www.ft.com/cms/s/2/504285c4-68b6-11da-bd30-0000779e2340,dwp_uuid=6f876a3c-e19f-11da-bf4c-0000779e2340.html). *Financial Times* (London). . Retrieved 19 March 2011.

[47] Protzman, Ferdinand (22 August 1989). "Westward Tide of East Germans Is a Popular No-Confidence Vote" (http://www.nytimes.com/1989/08/22/world/westward-tide-of-east-germans-is-a-popular-no-confidence-vote.html?pagewanted=all&src=pm). *www.nytimes.com* (The New York Times). . Retrieved 30 October 2011. "Behind the mass flight, Western experts say, is widespread and deepening disillusionment with the Honecker leadership's policies, particularly the refusal to consider the type of economic and political changes taking place elsewhere in Eastern Europe."

[48] "Gesetz zur Umsetzung des Beschlusses des Deutschen Bundestages vom 20. Juni 1991 zur Vollendung der Einheit Deutschlands" (http://bundesrecht.juris.de/berlin_bonng/index.html) (in German). Bundesministerium der Justiz. 26 April 1994. . Retrieved 19 April 2011.

[49] "Brennpunkt: Hauptstadt-Umzug" (http://www.focus.de/panorama/boulevard/brennpunkt-hauptstadt-umzug_aid_175751.html) (in German). *Focus* (Munich). 12 April 1999. . Retrieved 19 March 2011.

[50] Dempsey, Judy (31 October 2006). "Germany is planning a Bosnia withdrawal" (http://www.nytimes.com/2006/10/31/world/europe/31iht-germany.3343963.html). *International Herald Tribune* (Paris). . Retrieved 7 May 2011.

[51] Merz, Sebastian (November 2007). "Still on the way to Afghanistan? Germany and its forces in the Hindu Kush" (http://www.sipri.org/research/conflict/publications/merz) (PDF). Stockholm International Peace Research Institute. pp. 2, 3. . Retrieved 16 April 2011.

[52] "World Factbook" (https://www.cia.gov/library/publications/the-world-factbook/geos/gm.html). CIA. . Retrieved 26 March 2011.

[53] "Climate in Germany" (http://www.germanculture.com.ua/library/facts/bl_climate.htm). GermanCulture. . Retrieved 26 March 2011.

[54] "Terrestrial Ecoregions" (http://wwf.panda.org/about_our_earth/ecoregions/ecoregion_list/). WWF. . Retrieved 19 March 2011.

[55] Strohm, Kathrin (May 2010). "Arable farming in Germany" (http://www.agribenchmark.org/fileadmin/freefiles/ccc_2010/Poster_Germany.pdf). Agri benchmark. . Retrieved 14 April 2011.

[56] Bekker, Henk (2005). *Adventure Guide Germany*. Hunter. p. 14. ISBN 9781588435033.

[57] "Zoo Facts" (http://www.americanzoos.info/Zoofacts.html). Zoos and Aquariums of America. . Retrieved 16 April 2011.

[58] "Der Zoologische Garten Berlin" (http://www.zoo-berlin.de/zoo/unternehmen/historie.html) (in German). Zoo Berlin. . Retrieved 19 March 2011.

[59] "Basic Law for the Federal Republic of Germany" (https://www.btg-bestellservice.de/pdf/80201000.pdf). *Deutscher Bundestag*. Btg-bestellservice. October 2010. . Retrieved 14 April 2011.

[60] "Christian Democratic Union/Christian Social Union" (http://countrystudies.us/germany/159.htm). U.S. Library of Congress. . Retrieved 26 March 2011.

[61] "Federal Constitutional Court" (http://www.bundesverfassungsgericht.de/en/index.html). Bundesverfassungsgericht. . Retrieved 26 March 2011.

[62] "Völkerstrafgesetz Teil 1 Allgemeine Regelungen" (http://bundesrecht.juris.de/vstgb/BJNR225410002.html#BJNR225410002BJNG000200000) (in German). Bundesministerium der Justiz. . Retrieved 19 April 2011.

[63] "§ 2 Strafvollzugsgesetz" (http://www.gesetze-im-internet.de/stvollzg/__2.html) (in German). Bundesministerium der Justiz. . Retrieved 26 March 2011.

[64] Jehle, Jörg-Martin; German Federal Ministry of Justice (2009). *Criminal Justice in Germany* (http://books.google.com/books?id=-V-ng-8jOoQC&pg=PA23). Forum-Verlag. p. 23. ISBN 9783936999518. .

[65] Casper, Gerhard; Zeisel, Hans (January 1972). "Lay Judges in the German Criminal Courts". *Journal of Legal Studies* **1** (1): 141. JSTOR 724014.

[66] The individual denomination is either *Land* [state], *Freistaat* [free state] or *Freie (und) Hansestadt* [free (and) Hanseatic city].
"The Federal States" (http://www.bundesrat.de/nn_11006/EN/organisation-en/laender-en/laender-en-node.html?__nnn=true). *www.bundesrat.de*. Bundesrat of Germany. . Retrieved 17 July 2011.
"Amtliche Bezeichnung der Bundesländer [Official denomination of federated states]" (http://www.auswaertiges-amt.de/DE/Infoservice/Terminologie/Bundeslaender/Uebersicht_node.html) (in German) (PDF; download file "Englisch"). *www.auswaertiges-amt.de*. Federal Foreign Office. . Retrieved 22 October 2011.

[67] "Example for state constitution: "Constitution of the Land of North Rhine-Westphalia"" (http://www.landtag.nrw.de/portal/WWW/GB_I/I.7/Europa/Wissenswertes/English_information/North_Rhine_Westphalia_Constitution_revised.jsp). Landtag (state assembly) of North Rhine-Westphalia. . Retrieved 17 July 2011.

[68] "Kreisfreie Städte und Landkreise nach Fläche und Bevölkerung 31.12.2010" (https://www.destatis.de/DE/ZahlenFakten/LaenderRegionen/Regionales/Gemeindeverzeichnis/Administrativ/Aktuell/04Kreise.xls;jsessionid=611904454C9287BBBD47BE5D6103CAC1.cae2?__blob=publicationFile) (in German) (XLS). Statistisches Bundesamt Deutschland. October 2011. . Retrieved 6 April 2012.

[69] "German Missions Abroad" (http://www.auswaertiges-amt.de/EN/AAmt/Auslandsvertretungen/Uebersicht_node.html). German Federal Foreign Office. . Retrieved 26 March 2011.

[70] "The EU budget 2011 in figures" (http://ec.europa.eu/budget/figures/2011/2011_en.cfm). European Commission. . Retrieved 6 May 2011.

[71] "United Nations regular budget for the year 2011" (http://www.un.org/ga/search/view_doc.asp?symbol=ST/ADM/SER.B/824). UN Committee on Contributions. . Retrieved 6 May 2011.

[72] "Declaration by the Franco-German Defence and Security Council" (http://www.ambafrance-uk.org/Declaration-by-the-Franco-German,4519.html). French Embassy UK. 13 May 2004. . Retrieved 19 March 2011.

[73] Freed, John C. (4 April 2008). "The leader of Europe? Answers an ocean apart" (http://www.nytimes.com/2008/04/04/world/europe/04iht-poll.4.11666423.html). *The New York Times*. . Retrieved 28 March 2011.

[74] "Aims of German development policy" (http://www.bmz.de/en/index.html). Federal Ministry for Economic Cooperation and Development. 10 April 2008. . Retrieved 26 March 2011.

[75] "Net Official Development Assistance 2009" (http://www.oecd.org/dataoecd/17/9/44981892.pdf). OECD. . Retrieved 26 March 2011.

[76] "Speech by Chancellor Angela Merkel to the United Nations General Assembly" (http://www.bundesregierung.de/nn_6566/Content/EN/Reden/2010/2010-09-21-merkel-mdg-gipfel.html). *Die Bundesregierung*. 21 September 2010. . Retrieved 18 March 2011.

[77] Harrison, Hope (2004). "American détente and German ostpolitik, 1969–1972" (http://www.ghi-dc.org/files/publications/bu_supp/supp1/supp-01_005.pdf). *Bulletin Supplement* (German Historical Institute) **1**. . Retrieved 26 March 2011.

[78] "Germany's New Face Abroad" (http://www.dw-world.de/dw/article/0,2144,1741310,00.html). *Deutsche Welle*. 14 October 2005. . Retrieved 26 March 2011.

[79] "Ready for a Bush hug?" (http://www.economist.com/node/7141311?story_id=7141311). *The Economist*. 6 July 2006. . Retrieved 19 March 2011.

[80] "U.S.-German Economic Relations Factsheet" (http://germany.usembassy.gov/germany/img/assets/9336/econ_factsheet_may2006.pdf). U.S. Embassy in Berlin. May 2006. . Retrieved 26 March 2011.

[81] "Background paper on SIPRI military expenditure data, 2010" (http://www.sipri.org/research/armaments/milex/factsheet2010). Stockholm International Peace Research Institute. 11 April 2011. . Retrieved 7 May 2011.

[82] "Grundgesetz für die Bundesrepublik Deutschland, Artikel 65a,87,115b" (http://www.gesetze-im-internet.de/bundesrecht/gg/gesamt.pdf) (in German). Bundesministerium der Justiz. . Retrieved 19 March 2011.

[83] "Die Stärke der Streitkräfte" (http://www.bundeswehr.de/portal/a/bwde/!ut/p/c4/DcmxDYAwDATAWVgg7unYAugc8kSWI4OMIesTXXm002D8SeWQy7jRStshc-4p94L0hENCnXEGUvXXSuMKG8FwBd26TD9uIZiT/) (in German). Bundeswehr. . Retrieved 5 June 2011.

[84] "Ausblick: Die Bundeswehr der Zukunft" (http://www.bundeswehr.de/portal/a/bwde/!ut/p/c4/04_SB8K8xLLM9MSSzPy8xBz9CP3I5EyrpHK9pPKUVL3ikqLUzJLsosTUtJJUvbzU0vTU4pLEnJLSvHRUuYKcxDygoH5BtqMiAMTJdF8!/) (in German). Bundeswehr. . Retrieved 5 June 2011.

[85] "Einsatzzahlen – Die Stärke der deutschen Einsatzkontingente" (http://www.bundeswehr.de/portal/a/bwde/einsaetze/einsatzzahlen?yw_contentURL=/C1256EF4002AED30/W264VFT2439INFODE/content.jsp) (in German). Bundeswehr. . Retrieved 14 April 2011.

[86] Connolly, Kate (22 November 2010). "Germany to abolish compulsory military service" (http://www.guardian.co.uk/world/2010/nov/22/germany-abolish-compulsory-military-service). *The Guardian* (UK). . Retrieved 7 April 2011.

[87] Pidd, Helen (16 March 2011). "Marching orders for conscription in Germany, but what will take its place?" (http://www.guardian.co.uk/world/2011/mar/16/conscription-germany-army). *The Guardian* (UK). . Retrieved 7 April 2011.

[88] "Frauen in der Bundeswehr" (http://www.bundeswehr.de/portal/a/bwde/!ut/p/c4/FcwxEoUgDAXAE0l6O0_x1YZ5QMSMEp2In-urs_3STC_FXzKqHIqdRpqi9KG50BK7qxpL3Qy8VHbZbk07MqtbDDerF_WJzYdGv286DbmAJj26iLgynaUM

) (in German). Bundeswehr. . Retrieved 14 April 2011.

[89] Norris, Floyd (20 February 2010). "A Shift in the Export Powerhouses" (http://www.nytimes.com/2010/02/20/business/economy/20charts.html). *The New York Times*. . Retrieved 27 March 2011.

[90] "CPI 2009 table" (http://www.transparency.org/policy_research/surveys_indices/cpi/2009/cpi_2009_table). Transparency International. . Retrieved 27 March 2011.

[91] "The Innovation Imperative in Manufacturing: How the United States Can Restore Its Edge" (http://www.bcg.com/documents/file15445.pdf). Boston Consulting Group. March 2009. . Retrieved 19 March 2011.

[92] "Gross domestic product (2009)" (http://siteresources.worldbank.org/DATASTATISTICS/Resources/GDP.pdf). *The World Bank: World Development Indicators database*. World Bank. 27 September 2010. . Retrieved 1 January 2011.
Field listing – GDP (official exchange rate) (https://www.cia.gov/library/publications/the-world-factbook/fields/2195.html), CIA World Factbook

[93] "The Economic Complexity Observatory | Macro Connections" (http://macroconnections.media.mit.edu/featured/economic-complexity-observatory/). Macroconnections.media.mit.edu. . Retrieved 2012-04-05.

[94] "Wind Power" (http://web.archive.org/web/20061210163253/http://www.german-renewable-energy.com/Renewables/Navigation/Englisch/wind-power.html). Federal Ministry of Economics and Technology. Archived from the original (http://www.german-renewable-energy.com/Renewables/Navigation/Englisch/wind-power.html) on 2006-12-10. . Retrieved 27 March 2011.

[95] UFI, the Global Association of the Exhibition Industry (2008). "Euro Fair Statistics 2008" (http://www.auma.de/_pages/d/16_Download/download/FKM/EuroFairStatistics_2008.pdf). AUMA Ausstellungs- und Messe-Ausschuss der Deutschen Wirtschaft e.V.. p. 12. . Retrieved 24 September 2011.

[96] Andrews, Edmund L. (1 January 2002). "Germans Say Goodbye to the Mark, a Symbol of Strength and Unity" (http://www.nytimes.com/2002/01/01/world/germans-say-goodbye-to-the-mark-a-symbol-of-strength-and-unity.html). *The New York Times*. . Retrieved 18 March 2011.

[97] Taylor Martin, Susan (28 December 1998). "On Jan. 1, out of many arises one Euro". *St. Petersburg Times*: p. National, 1.A.

[98] Berg, S.; Winter, S.; Wassermann, A. (5 September 2005). "The Price of a Failed Reunification" (http://www.spiegel.de/international/spiegel/0,1518,373639,00.html). *Spiegel Online*. . Retrieved 28 November 2006.

[99] Kulish, Nicholas (19 June 2009). "In East Germany, a Decline as Stark as a Wall" (http://www.nytimes.com/2009/06/19/world/europe/19germany.html). *The New York Times*. . Retrieved 27 March 2011.

[100] "Germany agrees on 50-billion-euro stimulus plan" (http://www.france24.com/en/20090106-germany-agrees-new-50-billion-euro-stimulus-plan). *France 24*. 6 January 2009. . Retrieved 27 March 2011.

[101] "The 100 Top Brands 2010" (http://www.interbrand.com/en/best-global-brands/best-global-brands-2008/best-global-brands-2010.aspx). Interbrand. . Retrieved 27 March 2011.

[102] Gavin, Mike (23 September 2010). "Germany Has 1,000 Market-Leading Companies, Manager-Magazin Says" (http://www.businessweek.com/news/2010-09-23/germany-has-1-000-market-leading-companies-manager-magazin-says.html). *Businessweek* (New York). . Retrieved 27 March 2011.

[103] "Global 500: Countries – Germany" (http://money.cnn.com/magazines/fortune/global500/2010/countries/Germany.html). *Forbes*. 26 July 2010. . Retrieved 27 March 2011.

[104] "Autobahn-Temporegelung" (http://www.presse.adac.de/standpunkte/Verkehr/Autobahn_Temporegelung.asp?active1=tcm:11-18784-4) (in German) (Press release). ADAC. June 2010. . Retrieved 19 March 2011.

[105] "Geschäftsbericht 2006" (http://web.archive.org/web/20070809140315/http://www.db.de/site/bahn/de/unternehmen/investor__relations/finanzberichte/geschaeftsbericht/geschaeftsbericht__2006.html) (in German). Deutsche Bahn. Archived from the original (http://www.db.de/site/bahn/de/unternehmen/investor__relations/finanzberichte/geschaeftsbericht/geschaeftsbericht__2006.html) on 2007-08-09. . Retrieved 27 March 2011.

[106] "Airports in Germany" (http://www.aircraft-charter-world.com/airports/europe/germany.htm). Air Broker Center International. . Retrieved 16 April 2011.

[107] "Overview/Data: Germany" (http://www.eia.gov/countries/country-data.cfm?fips=GM). U.S. Energy Information Administration. 30 June 2010. . Retrieved 19 April 2011.

[108] "Energy imports, net (% of energy use)" (http://data.worldbank.org/indicator/EG.IMP.CONS.ZS). The World Bank Group. . Retrieved 18 April 2011.

[109] Reuters Berlin (7 July 2010). "* Environment * Renewable energy Germany targets switch to 100% renewables for its electricity by 2050" (http://www.guardian.co.uk/environment/2010/jul/07/germany-renewable-energy-electricity). *The Guardian* (UK). . Retrieved 18 April 2011.

[110] "Primärenergieverbrauch nach Energieträgern" (http://www.bmwi.de/BMWi/Redaktion/Binaer/Energiedaten/energiegewinnung-und-energieverbrauch2-primaerenergieverbrauch.xls) (in German). Bundesministerium für Wirtschaft und Technologie. December 2010. . Retrieved 18 April 2011.

[111] "Germany split over green energy" (http://news.bbc.co.uk/2/hi/europe/4295389.stm). *BBC News*. 25 February 2005. . Retrieved 27 March 2011.

[112] "Deutschland erfüllte 2008 seine Klimaschutzverpflichtung nach dem Kyoto-Protokoll" (http://www.kst.portalu.de/dokumente/Aktuelles/Inventar-Bericht_Treibhausgas_2008.pdf) (in German) (Press release). Umweltbundesamt. 1 February 2010. . Retrieved 19 March 2011.

[113] "Germany greenest country in the world" (http://articles.timesofindia.indiatimes.com/2008-06-21/pollution/27752791_1_energy-security-global-energy-germany). *The Times of India* (New Delhi). 21 June 2008. . Retrieved 26 March 2011.

[114] "Treibhausgas-Emissionen der Europäischen Union ..." (http://www.umweltbundesamt-daten-zur-umwelt.de/umweltdaten/public/document/downloadImage.do?ident=18198) (in German) (PDF). Umweltbundesamt. 2009. . Retrieved 10 May 2011.

[115] "Germany's Technological Performance" (http://www.bmbf.de/en/1869.php). Federal Ministry of Education and Research. 11 January 2007. . Retrieved 21 August 2011.

[116] "Nobel Prize" (http://nobelprize.org/). Nobelprize.org. . Retrieved 27 March 2011.

[117] "Swedish academy awards" (http://www.sciencenews.org/view/generic/id/63944/title/Swedish_academy_awards). *ScienceNews*. . Retrieved 1 October 2010.

[118] National Science Nobel Prize shares 1901–2009 by citizenship at the time of the award (http://www.idsia.ch/~juergen/sci.html) and by country of birth (http://www.idsia.ch/~juergen/scinat.html). From Schmidhuber, J. (2010). "Evolution of National Nobel Prize Shares in the 20th century" (http://www.idsia.ch/~juergen/nobelshare.html). . Retrieved 27 March 2011.

[119] Roberts, J. M. (2002). *The New Penguin History of the World*. Allen Lane. p. 1014. ISBN 9780713996111.

[120] "The First Nobel Prize" (http://www.dw-world.de/dw/article/0,,5984670,00.html). *Deutsche Welle*. 8 September 2010. . Retrieved 27 March 2011.

[121] "Gottfried Wilhelm Leibniz Prize" (http://web.archive.org/web/20080621091621/http://www.dfg.de/en/research_funding/scientific_prizes/gw_leibniz_prize.html). DFG. Archived from the original (http://www.dfg.de/en/research_funding/scientific_prizes/gw_leibniz_prize.html) on 2008-06-21. . Retrieved 27 March 2011.

[122] Bianchi, Luigi. "The Great Electromechanical Computers" (http://www.yorku.ca/lbianchi/sts3700b/lecture17a.html). York University. . Retrieved 17 April 2011.

[123] "The Zeppelin" (http://www.centennialofflight.gov/essay/Lighter_than_air/zeppelin/LTA8.htm). U.S. Centennial of Flight Commission. . Retrieved 27 March 2011.

[124] "Historical figures in telecommunications" (http://www.itu.int/aboutitu/HistoricalFigures.html). International Telecommunication Union. 14 January 2004. . Retrieved 27 March 2011.

[125] Roland Berger Strategy Consultants: *Green Growth, Green Profit – How Green Transformation Boosts Business* Palgrave Macmillan, New York 2010, ISBN 978-0-230-28543-9

[126] "Country Comparison :: Population" (https://www.cia.gov/library/publications/the-world-factbook/rankorder/2119rank.html?countryName=Lebanon&countryCode=le®ionCode=me&rank=126). CIA. . Retrieved 26 June 2011.

[127] Destatis. "Durchschnittliche Kinderzahl 2008 in den neuen Ländern angestiegen" (http://web.archive.org/web/20101229095132/http://www.destatis.de/jetspeed/portal/cms/Sites/destatis/Internet/DE/Presse/pm/2010/01/PD10__034__12641,templateId=renderPrint.psml) (in German). Archived from the original (http://www.destatis.de/jetspeed/portal/cms/Sites/destatis/Internet/DE/Presse/pm/2010/01/PD10__034__12641,templateId=renderPrint.psml) on 29 December 2010. . Retrieved 28 March 2011.

[128] "Demographic Transition Model" (http://geographyfieldwork.com/DemographicTransition.htm). Barcelona Field Studies Centre. 27 September 2009. . Retrieved 28 March 2011.

[129] "Im Jahr 2060 wird jeder Siebente 80 Jahre oder älter sein" (https://www.destatis.de/DE/PresseService/Presse/Pressemitteilungen/2009/11/PD09_435_12411.html;jsessionid=7247D871DEB96E4AE313EE55A45CEFC1.cae1) (in German) (Press release). Destatis. 18 November 2009. . Retrieved 6 April 2012.
Details on the methodology, detailed tables, etc. are provided at "Bevölkerungsentwicklung in Deutschland bis 2060" (http://web.archive.org/web/20101229190016/http://www.destatis.de/jetspeed/portal/cms/Sites/destatis/Internet/DE/Presse/abisz/Bevoelkerungsvorausberechnung,templateId=renderPrint.psml) (in German). Statistisches Bundesamt. 18 November 2009. Archived from the original (http://www.destatis.de/jetspeed/portal/cms/Sites/destatis/Internet/DE/Presse/abisz/Bevoelkerungsvorausberechnung,templateId=renderPrint.psml) on 29 December 2010. . Retrieved 15 February 2011.

[130] (in German) (PDF) *Bevölkerung und Erwerbstätigkeit:Bevölkerung mit Migrationshintergrund – Ergebnisse des Mikrozensus 2010 [Population and employment: Population with migrant background – Results of the 2010 microcensus]* (https://www.destatis.de/DE/Publikationen/Thematisch/Bevoelkerung/MigrationIntegration/Migrationshintergrund2010220107004.pdf?__blob=publicationFile). Fachserie 1 Reihe 2.2. Statistisches Bundesamt. 14 July 2010. pp. 6–8. Artikelnummer: 2010220107004. . Retrieved 7 March 2012.
The Federal Statistical Office defines persons with a migrant background as all persons who migrated to the present area of the Federal Republic of Germany after 1949, plus all foreign nationals born in Germany and all persons born in Germany as German nationals with at least one parent who migrated to Germany or was born in Germany as a foreign national.

[131] "International Migration 2006" (http://www.un.org/esa/population/publications/2006Migration_Chart/Migration2006.pdf). UN Department of Economic and Social Affairs. . Retrieved 18 March 2011.

[132] "Germany" (http://www.focus-migration.de/Germany.1509.0.html?&L=1). Focus-Migration. . Retrieved 28 March 2011.

[133] "20% of Germans have immigrant roots". *Burlington Free Press*: p. 4A. 15 July 2010.

[134] "Bevölkerung nach Migrationshintergrund" (http://web.archive.org/web/20101229095012/http://www.destatis.de/jetspeed/portal/cms/Sites/destatis/Internet/DE/Content/Statistiken/Bevoelkerung/MigrationIntegration/Migrationshintergrund/Tabellen/Content100/MigrationshintergrundStaatsangehoerigkeit,templateId=renderPrint.psml) (in German). German Federal Statistical Office. Archived from the original (http://www.destatis.de/jetspeed/portal/cms/Sites/destatis/Internet/DE/Content/Statistiken/Bevoelkerung/MigrationIntegration/Migrationshintergrund/Tabellen/Content100/MigrationshintergrundStaatsangehoerigkeit,templateId=renderPrint.psml) on 29 December 2010. . Retrieved 28 March 2011.

[135] "Fewer Ethnic Germans Immigrating to Ancestral Homeland" (http://www.migrationinformation.org/Feature/display.cfm?ID=201). Migration Information Source. February 2004. . Retrieved 28 March 2011.

[136] "Regionales Monitoring 2008 – Daten und Karten zu den Europäischen Metropolregionen (EMR) in Deutschland" (http://www.deutsche-metropolregionen.org/fileadmin/ikm/IKM-Veroeffentlichungen/IKM-Monitoring2008_lite.pdf) (in German). Bundesamt für Bauwesen und Raumordnung. 2008. p. 10. . Retrieved 11 June 2011.

[137] "EKD-Statistik: Christen in Deutschland 2007" (http://www.ekd.de/statistik/mitglieder.html) (in German). Evangelische Kirche in Deutschland. . Retrieved 28 March 2011.

[138] "Konfessionen in Deutschland" (http://fowid.de/fileadmin/datenarchiv/Religionszugehoerigkeit_Bevoelkerung__1950-2008.pdf) (in German). Fowid. 9 September 2009. . Retrieved 28 March 2011.

[139] "Chapter 2: Wie viele Muslime leben in Deutschland?" (http://www.bmi.bund.de/cae/servlet/contentblob/566008/publicationFile/31710/vollversion_studie_muslim_leben_deutschland_.pdf;jsessionid=6B8CD26E2AC179111AF4F75650B84B1A) (in German) (PDF). *Muslimisches Leben in Deutschland*. Bundesamt für Migration und Flüchtlinge. June 2009. pp. 80, 97. ISBN 978-3-9812115-1-1. . Retrieved 28 March 2011.

[140] "Religionen in Deutschland: Mitgliederzahlen" (http://www.remid.de/remid_info_zahlen.htm) (in German). Religionswissenschaftlicher Medien- und Informationsdienst. 31 October 2009. . Retrieved 28 March 2011.

[141] Blake, Mariah (10 November 2006). "In Nazi cradle, Germany marks Jewish renaissance" (http://www.csmonitor.com/2006/1110/p25s02-woeu.html). *Christian Science Monitor*. . Retrieved 28 March 2011.

[142] Schnabel, U. (15 March 2007). "Buddhismus Eine Religion ohne Gott" (http://www.zeit.de/2007/12/Buddhismus) (in German). *Die Zeit* (Hamburg). . Retrieved 19 March 2011.

[143] European Commission (2006). "Special Eurobarometer 243: Europeans and their Languages (Survey)" (http://ec.europa.eu/public_opinion/archives/ebs/ebs_243_en.pdf). Europa (web portal). . Retrieved 28 March 2011.
European Commission (2006). "Special Eurobarometer 243: Europeans and their Languages (Executive Summary)" (http://ec.europa.eu/public_opinion/archives/ebs/ebs_243_sum_en.pdf). Europa (web portal). . Retrieved 28 March 2011.

[144] European Commission (2004). "Many tongues, one family. Languages in the European Union" (http://ec.europa.eu/publications/booklets/move/45/en.pdf). Europa (web portal). . Retrieved 28 March 2011.

[145] "Sprechen Sie Deutsch?" (http://www.economist.com/node/15731354). *The Economist*. 18 March 2010. . Retrieved 16 April 2011.

[146] Grotlüschen, Anke; Riekmann, Wibke (2011). "leo.- Level One Survey. Presseheft" (http://www.vhs-sachsen.de/uploads/media/Executive_Summary_leo_v8_Journalismusfassung.pdf) (in German). University of Hamburg. . Retrieved 15 April 2011.

[147] "Country profile: Germany" (http://lcweb2.loc.gov/frd/cs/profiles/Germany.pdf). Library of Congress. April 2008. . Retrieved 28 March 2011.

[148] "The Educational System in Germany" (http://academic.cuesta.edu/intlang/german/education.html). Cuesta College. 31 August 2002. . Retrieved 16 May 2011.

[149] "Top 100 World Universities" (http://web.archive.org/web/20080822124509/http://www.arwu.org/rank2008/ARWU2008_A(EN).htm). Academic Ranking of World Universities. Archived from the original (http://www.arwu.org/rank2008/ARWU2008_A(EN).htm) on 2008-08-22. . Retrieved 28 March 2011.

[150] "Tuition Fees at university in Germany" (http://www.studyineurope.eu/study-in-germany/admission/tuition-fees). StudyinEurope.eu. 2009. . Retrieved 19 March 2011.

[151] *Health Care Systems in Transition: Germany* (http://www.euro.who.int/__data/assets/pdf_file/0010/80776/E68952.pdf). European Observatory on Health Care Systems. 2000. p. 8. AMS 5012667 (DEU). . Retrieved 15 April 2011.

[152] "Die Gesundheitsreform 2007: Was hat sich geändert?" (http://www.krankenkassen.de/gesetzliche-krankenkassen/gesundheitsreform/Gesundheitsreform-zentrale-Punkte-Fragen/). Krankenkassen.de. . Retrieved 16 April 2012.

[153] "Core Health Indicators" (http://apps.who.int/whosis/database/core/core_select.cfm). World Health Organization. . Retrieved 6 June 2011.

[154] "2010: Herz-/Kreislauferkrankungen verursachen 41 % aller Todesfälle" (https://www.destatis.de/DE/ZahlenFakten/GesellschaftStaat/Gesundheit/Todesursachen/Aktuell.html;jsessionid=B6BD78C7EDB6FE6693DE58EA7B23D148.cae1) (in German). Destatis.de. . Retrieved 6 April 2012.

[155] "Country Profile Germany" (http://lcweb2.loc.gov/frd/cs/profiles/Germany.pdf) (PDF). Library of Congress Federal Research Division. April 2008. . Retrieved 7 May 2011.
This article may incorporate text from this source, which is in the public domain.

[156] Wasser, Jeremy (6 April 2006). "Spätzle Westerns" (http://www.spiegel.de/international/0,1518,410135,00.html). *Spiegel Online International*. . Retrieved 28 March 2011.

[157] "Unbelievable Multitude" (http://www.dw-world.de/dw/0,,8009,00.html). *Deutsche Welle*. . Retrieved 28 March 2011.

[158] "World Heritage Sites in Germany" (http://www.worldheritagesite.org/countries/germany.html). UNESCO. . Retrieved 3 October 2010.

[159] "Human Development Report 2010 Table 4 Gender Inequality Index" (http://hdr.undp.org/en/media/HDR_2010_EN_Tables_reprint.pdf). United Nations Development Programme. pp. 156–160. . Retrieved 20 April 2011.

[160] "Germany extends gay rights" (http://www.news24.com/World/News/Germany-extends-gay-rights-20041029). News24. 29 October 2004. . Retrieved 19 March 2011.

[161] Heckmann, Friedrich (2003). *The Integration of Immigrants in European Societies: national differences and trends of convergence*. Lucius & Lucius. pp. 51 ff. ISBN 978-3-8282-0181-1.

[162] "2010 Anholt-GfK Roper Nation Brands Index" (http://www.gfk.com/group/press_information/press_releases/006688/index.en.html) (Press release). GfK. 12 October 2010. . Retrieved 28 March 2011.

[163] "Views of US Continue to Improve in 2011 BBC Country Rating Poll" (http://www.worldpublicopinion.org/pipa/articles/views_on_countriesregions_bt/680.php?nid=&id=&pnt=680&lb=#ger). *Worldpublicopinion.org*. 7 March 2011. . Retrieved 28 March 2011.

[164] *A Dictionary of Architecture and Landscape Architecture*. Oxford University Press. 2006. p. 880. ISBN 0-19-860678-8.

[165] "Germany's flailing music industry seeks new talent" (http://www.dw-world.de/dw/article/0,,5200389,00.html). *Deutsche Welle*. 2 February 2010. . Retrieved 28 March 2011.

[166] Espmark, Kjell (3 December 1999). "The Nobel Prize in Literature" (http://nobelprize.org/nobel_prizes/literature/articles/espmark/index.html). Nobelprize.org. . Retrieved 28 March 2011.

[167] "Land of ideas" (http://land-der-ideen.matrix.de/CDA/facts_printing,4563,0,,en.html). Land-der-ideen.matrix.de. . Retrieved 19 March 2011.

[168] Weidhaas, Peter; Gossage, Carolyn; Wright, Wendy A. (2007). *A History of the Frankfurt Book Fair*. Dundurn Press Ltd.. pp. 11 ff. ISBN 9781550027440.

[169] Searle, John (1987). "Introduction". *The Blackwell Companion to Philosophy*. Wiley-Blackwell.

[170] Bordwell, David; Thompson, Kristin (2003) [1994]. "The Introduction of Sound". *Film History: An Introduction* (2nd ed.). McGraw-Hill. p. 204. ISBN 978-0-07-115141-2.

[171] "Rainer Werner Fassbinder" (http://www.fassbinderfoundation.de/node.php/en/home). Fassbinder Foundation. . Retrieved 28 March 2011.

[172] "2006 FIAPF accredited Festivals Directory" (http://www.fiapf.org/pdf/2006accreditedFestivalsDirectory.pdf). International Federation of Film Producers Associations. . Retrieved 28 March 2011.

[173] "Awards:Das Leben der Anderen" (http://www.imdb.com/title/tt0405094/awards). IMDb. . Retrieved 28 March 2011.

[174] "Country profile: Germany" (http://news.bbc.co.uk/2/hi/europe/country_profiles/1047864.stm). *BBC News*. . Retrieved 28 March 2011.

[175] "Guide to German Sausages & Meat Products" (http://www.germanfoods.org/consumer/facts/guidetosausages.cfm). German Foods North America. . Retrieved 6 April 2012.

[176] "Germany Country Profiles for Organic Agriculture" (http://www.fao.org/organicag/display/work/display_2.asp?country=DEU&lang=en&disp=summaries). Food and Agriculture Organization. . Retrieved 6 May 2011.

[177] Schneibel, Gerhard (23 April 2010). "Brewers not worried by beer consumption drop" (http://www.dw.de/dw/article/0,,5489225,00.html). *Deutsche Welle* (Bonn). . Retrieved 6 April 2012.

[178] "Schnitzel Outcooks Spaghetti in Michelin Guide" (http://www.dw.de/dw/article/0,,2914502_page_0,00.html). *Deutsche Welle* (Bonn). 15 November 2007. . Retrieved 6 April 2012.

[179] "German cuisine beats Italy, Spain in gourmet stars" (http://in.reuters.com/article/2007/11/14/us-germany-food-idINL1447732320071114). *Reuters*. 28 March 2011. . Retrieved 19 March 2011.

[180] "Germany Info: Culture & Life: Sports" (http://www.germany.info/relaunch/culture/life/sports.html). Germany Embassy in Washington, D.C. . Retrieved 28 March 2011.

[181] Ornstein, David (23 October 2006). "What we will miss about Michael Schumacher" (http://www.guardian.co.uk/sport/2006/oct/23/formulaone.sport). *The Guardian* (UK). . Retrieved 19 March 2011.

[182] "Beijing 2008 Medal Table" (http://www.olympic.org/medallists-results?athletename=&category=343488&games=1333952&sport=&event=&mengender=false&womengender=false&mixedgender=false&teamclassification=false&individualclassification=false&continent=&country=&goldmedal=true&silvermedal=true&bronzemedal=true&worldrecord=false&olympicrecord=false&targetresults=true). International Olympic Committee. . Retrieved 19 March 2011.

[183] "Turin 2006 Medal Table" (http://www.olympic.org/medallists-results?athletename=&category=343486&games=1334152&sport=&event=&mengender=false&womengender=false&mixedgender=false&teamclassification=false&individualclassification=false&continent=&country=&goldmedal=true&silvermedal=true&bronzemedal=true&worldrecord=false&olympicrecord=false&targetresults=true&sortorder=medal&sortorder=country). International Olympic Committee. . Retrieved 19 March 2011.

Work cited

Fulbrook, Mary (1991). *A Concise History of Germany*. Cambridge University Press. ISBN 9780521368360.

External links

- deutschland.de (http://www.deutschland.de/en/home-page.html?tx_fdfxyaml_pi1) – Official Germany portal (non-profit)
- Official site of German Chancellor (http://www.bundeskanzlerin.de/Webs/BK/En/Homepage/home.html)
- Wikimedia Atlas of Germany
- Deutsche Welle (http://www.dw-world.de/) – Germany's international broadcaster
- Germany (http://ucblibraries.colorado.edu/govpubs/for/germany.htm) at *UCB Libraries GovPubs*
- Germany (http://www.dmoz.org/Regional/Europe/Germany/) at the Open Directory Project
- Facts about Germany (http://www.tatsachen-ueber-deutschland.de/en/) – by the German Federal Foreign Office
- Destatis.de (https://www.destatis.de/EN/Homepage.html) – Federal Statistical Office Germany
- Geographic data related to Germany (http://www.openstreetmap.org/browse/relation/51477) at OpenStreetMap
- Key Development Forecasts for Germany (http://www.ifs.du.edu/ifs/frm_CountryProfile.aspx?Country=DE) from International Futures

Article Sources and Contributors

Wittenberge *Source*: http://en.wikipedia.org/w/index.php?title=Wittenberge *Contributors*: Ahuskay, Cepheiden, DerBorg, Gettingtoit, Hughstew, Kwamikagami, Lschulz, Markussep, N-true, Ocolon, RJFJR, Rich Farmbrough, Sebastian scha., Segv11, Slambo, SluggoOne, Stemonitis, Timeineurope, Traveler100, Veledan, 11 anonymous edits

Westerrönfeld *Source*: http://en.wikipedia.org/w/index.php?title=Westerr%C3%B6nfeld *Contributors*: Ksnow, Wikidudeman

Municipalities of Germany *Source*: http://en.wikipedia.org/w/index.php?title=Municipalities_of_Germany *Contributors*: Acntx, Agathoclea, Carlossuarez46, Chanheigeorge, David Kernow, DerBorg, Ehrlich91, Kdammers, Mschiffler, Poulpy, Reinyday, Roke, Sardanaphalus, Stemonitis, Sundar1, Tobias Conradi, Valentinian, Worobiew, Yngvadottir, 3 anonymous edits

Gemeinde *Source*: http://en.wikipedia.org/w/index.php?title=Gemeinde *Contributors*: Borgx, Chanheigeorge, CharlotteWebb, DerBorg, Dirknachbar, F.chiodo, Grafen, IZAK, Jurema Oliveira, Mai-Sachme, Markussep, Moonraker12, Nyttend, PaulGarner, R9tgokunks, Radagast83, Sp4rr0W, TheMoth&TheButterfly, Tobias Conradi, TurkChan, Wikid77, Woohookitty, Xtv, Yakoo, 10 anonymous edits

Regierungsbezirk *Source*: http://en.wikipedia.org/w/index.php?title=Regierungsbezirk *Contributors*: (, Agathoclea, Ahoerstemeier, Andrew c, Angr, Badanedwa, Bermicourt, CalJW, Capitonym, Chanheigeorge, Crusoe8181, David Parker, Delirium, DerBorg, Djmutex, Docu, Donald Albury, Dvavasour, Ehrlich91, Elapsed, Eugene van der Pijll, Exec, Frokor, Funandtrvl, Gereonmc, GregorB, Ingolfson, JPosten, JeLuF, KaiKemmann, Kelisi, Kwamikagami, Lfh, LilHelpa, Ludger1961, Marek69, MegA, Meursault2004, Mrmuk, Numbo3, Olessi, Picapica, Piotrus, Ratzer, Reywas92, Rje, Robofish, Rossami, Royboycrashfan, Saintswithin, Sandman, Timwi, Tom, Ulf Heinsohn, Unyoyega, Worobiew, Δ, 30 anonymous edits

Rendsburg-Eckernförde *Source*: http://en.wikipedia.org/w/index.php?title=Rendsburg-Eckernf%C3%B6rde *Contributors*: Agathoclea, Ahoerstemeier, Andres, Arne-, Baldhur, Bermicourt, Darwinek, GeorgHH, Ggpauly, Good Olfactory, Hede2000, Inwind, Markussep, Olessi, Rich Farmbrough, Rjwilmsi, Sandman, Sebastian scha., Tovk909, Valentinian, WikHead, Wwilly, Ztronix, 14 anonymous edits

Amt (country subdivision) *Source*: http://en.wikipedia.org/w/index.php?title=Amt_%28country_subdivision%29 *Contributors*: 32X, Agathoclea, Ahoerstemeier, AllKnowing, Baldhur, Big Adamsky, Bingobangobongoboo, Blah(de), Byrial, Chris.usnames, David Kernow, DerBorg, DragonflySixtyseven, Dycedarg, Eequor, Ehrlich91, Emil Kastberg, Emvee, Gidonb, Gustavo Siqueira, Hairy Dude, HeartofaDog, Jóhann Heiðar Árnason, Kamui99, Kelisi, Khazar, Lilac Soul, Markussep, Mevsfotw, Moonraker12, Net-net, Neutronrocks, Olivier, Peyerk, Poulpy, Quiddity, Qwfp, R9tgokunks, Remes, Reywas92, SET, Sardanaphalus, Scoo, Sfdan, SimonP, The Phoenix, The Transhumanist, ThomasPusch, Thue, Timwi, Tiraios-of-Characene, Tirelietirelei, Tobias Conradi, Tommycw1, TurkChan, Valentinian, Winhunter, Wwoods, Zserghei, °, 20 anonymous edits

Schleswig-Holstein *Source*: http://en.wikipedia.org/w/index.php?title=Schleswig-Holstein *Contributors*: 12.47, 19.168, 52 Pickup, AbdulRahiem, Ahoerstemeier, AjaxSmack, Aldux, AndrewRT, Anthony Appleyard, Arbor, Arno, Avb, Axeman89, Axt, Baldhur, Barticus88, Bermicourt, Bigbossfarin, Bleh999, Bob Burkhardt, Boctheow, Bomac, Boris Kaiser, Bornsommer, BrianHansen, Brighterorange, BrownHairedGirl, Buaidh, Bz2, Caltas, Cantiorix, Capricorn42, Chris the speller, CieloEstrellado, CohenTheBavarian, Conversion script, DaQuirin, Dan Koehl, David Liuzzo, David.Monniaux, Dbfirs, De728631, Der Eberswalder, Der Statistiker, DerBorg, Derek Ross, Dgies, Djmutex, Doco, Docu, Domino theory, DrTorstenHenning, Dukeofomnium, ESkog, Egern, Emersoni, Evdrneut, FMB, Favonian, FreplySpang, Funnyhat, Gaius Cornelius, Geoffrie, Gilgamesh, Givegains, Goldfishbutt, Good Olfactory, Graham87, H005, Hayden120, Heinrich L., Henning M, Her5ry, HerkusMonte, Hildegund, I know jesus, IRP, IZAK, Iamiyouareyou, Icairns, Igoldste, Inwind, Isababa7, J. 'mach' wust, J.delanoy, JHK, JeLuF, Jeltz, Jennavecia, John, Joseph Solis in Australia, Kbdank71, Kelisi, Kingjeff, Kingpin13, Kirimsa, Kito, Koavf, Kresspahl, Kusma, Kwamikagami, LA2, Laurascudder, LeVoyageur, Leandrod, Lear 21, Leibniz, Lfh, Life of Riley, LilHelpa, Littlealien182, Llywrch, Looxix, Lord Pistachio, LoveActresses, Luuxlatino2960, Marco Polo, Markussep, Marsal20, Meursault2004, Michael Hardy, MichaelTinkler, Misteror, Morven, Morwen, Mr Rookles, Mr.Scholari, Musicman501, Naddy, Nasz, Neutrality, Nev, Nico, Nikio, Oknazevad, Olaf Simons, Olessi, Olivier, Oreo Priest, Ospalh, OwenBlacker, PBS-AWB, Paine Ellsworth, Pentegamer, Piotrus, Plastikspork, Quailman, RadicalBender, Rcrowdy, Reyk, Rich Farmbrough, Romarin, Rosenzweig, Ruhrjung, Sabri76, Saddhiyama, Sandman, Sannse, Schipleon, Schneelocke, Shakescene, Snowdog, Solumeiras, Stadersand, Suisui, Sundostund, Template namespace initialisation script, Terfili, Thamis, The Thing That Should Not Be, Thryduulf, Thue, Tkynerd, Tourboy, Tridesch, Ulamm, Ulf Heinsohn, Ulric1313, Utergar, Uthlande, UweBayern, Valentinian, Veledan, Walden69, Wdfarmer, Wikid77, Wjhonson, Woohookitty, Writtenright, Wwwillly, Wzwz, Ztronix, €pa, 279 anonymous edits

Germany *Source*: http://en.wikipedia.org/w/index.php?title=Germany *Contributors*: -- April, -Majestic-, 000peter, 02pollajo, 03md, 06alifar, 119, 123kelsey123, 1297, 14Adrian, 159753, 16@r, 1Winston, 247balla34, 2812, 334a, 52 Pickup, 7huizen, 8ung3st, A Werewolf, A bit iffy, A little insignificant, A-giau, ABCD, ABlockedUser, AJackl, ALE!, ANNRC, Aaker, Aaron, Aaron Einstein, Aaron Schulz, Aaronlws, Aazn, Abe Lincoln, Abonazzi, Abrech, Academic Challenger, Academicigbo, Acalamari, Acanon, Achates, Adam Bishop, Adam Lutostański, Adashiel, Adcva, Adherent of the Enlightenment 10.0, AdmRose, Aeusoes1, Agathoclea, Ahoerstemeier, Ahrarara, Ahuskay, Aitias, Aivazovsky, AjaxSmack, Ajenterp, Akanemoto, Akhristov, Akkolon, Akradecki, Alai, Alan De Smet, Alanmak, Alarics, Alberto3166, Alex S, Alex Usoltsev, Alex1011, Alexandru Busa, Alexandru Stanoi, Alfirin, Alfonsomedina1, Algebraist, Algont, Alison, Alixus, All N Ever, Allgermein, Allmightyduck, AllyUnion, AlmightyClam, Alopex, Alphachimp, Alphasinus, Alphax, Altenmann, Alterrabe, Amakuru, Amber-W-12, Amcaja, AmiDaniel, Amoreno65, Anaki72, Ancheta Wis, Andre Engels, Andrew Levine, Andrewlp1991, Andries, Andrij Kursetsky, Andris, Andrwsc, Andy Marchbanks, Andyjsmith, Andypandy.UK, Anetode, Anger22, Angr, AniRaptor2001, Animum, Anjwalker, Ann O'nyme, Anna Lincoln, Anoldtreeok, AnonMoos, Anonymous from the 21th century, Antandrus, Anteeru, Anthony3, Anthonymendoza, Antidote, Antientropic, Antonio Basto, Ap, Apeloverage, Apoc2400, Apokrif, Appleseed, Apuman, Aqwis, Aranherunar, Arathjp, Arbero, Arindam.2011, Aris Katsaris, ArmadilloFromHell, Arnomane, Arnowaschk, Arsonal, Art LaPella, Artichoker, Arunnplus, Asams10, Asarelah, Asdfasdf1231234, Ashton1983, Askalan, Asnatu wiki, Asocall, AsusFan, Ataraxis1492, Athinaios, Aths, Atlant, Atlanticpuffin, Atlantik, Atoric, Atrix20, Attilios, Atwardow, AuburnPilot, Aude, Audiovideo, Aufbauten, Augustus Rookwood, Auntof6, Auric, Auriong, Autocracy, Avala, Avatarion, AvicAWB, Avicennasis, Avraham, Awolf002, AxG, Axegod12, Axel Phony, AxelBoldt, Axpde, Axt, Aye Carumba Fajita Pizza, Azate, Aznpryde, B1llycl3m3nt, BD2412, BGManofID, BJess, BRG, Babakexorramdin, Babbage, Backspace, Baldhur, Ballsontheroad, BalowStar, Bambuway, Banana04131, BanyanTree, Bapon, Barberio, Bardhylius, Barek, Baristarim, Barliner, Baronnet, Barras, Barryob, Barutazaru, Baseballnut290, Bastin, Battem, Baudrillard, Bazonka, Bbenjoe, Bbruchs, Bcorr, Beagel, Beetstra, Bejinhan, Beland, BenMerill, Bender235, Benefitov, BenjaT, Benne, Benster1510, Bento00, Benwildeboer, Benznine, Berkut, BernardaAlba, Bernardissimo, BernhardtP, Bert-25, Beta m, Betacommand, Betterworld, Bevo74, Bg007, Bhuck, BigBen212, Biker Biker, BilCat, Bilbo571, Bippy1990, Bjarki S, Bkell, Blablaaa, Blacha, BlackGothFaerie, Blackjack48, Blacklite, Blake24, Blake3522, BlazeTheMovieFan, Bleh999, Bletch, Blinder Seher, Blizzardstep0, Bloodofox, Blue-Haired Lawyer, Blue520, BlueMars, Bluemask, Bluezy, Blur4760, Bml, Bob rulz, Bob5409, Bob889900, Bobblewik, Bobby122, Bobet, Bobo192, Bocafan76, Bohuiginn, Boothy443, BorgHunter, Bornsommer, BoroughPeter, Boson, Bowman17, Boznia, Brad101, Bradeos Graphon, Brahmaputra, Brain40, Brainbark, Branddobbe, Brandmeister, Brandmeister (old), Brandt1, Breathe, BrendelSignature, Breno, Brian0918, Brighterorange, Brion VIBBER, Britannicus, BritishWatcher, Brockert, BrokenSegue, Brossow, Bruceknibbs, Brumski, Brutaldeluxe, Brutannica, Bryan Derksen, Bssc81, Buaidh, Bubba hotep, Buchanan-Hermit, Buddha24, BumHoleo, Bunny-chan, Burpelson AFB, Bushy moustache, BuzzWoof, Bwithh, ByrnedHead, CJ Aikman, CORNELIUSSEON, CSWarren, Cactus.man, Caerwine, Cakeandicecream, CalJW, CalendarWatcher, CalicoCatLover, Caltas, Calum MacÜisdean, Cam275, CambridgeBayWeather, Cambyses, Cameltoe237120236419324, Camilton, Can't sleep, clown will eat me, CanDo, Canadian-Bacon, Canaen, Canationalist, Canderra, Caniago, Canterbury Tail, Cantus, Captain Courageous, Captain Future, CaptainUnderpants, CaptainVindaloo, Captainspizzo, Carabinieri, Caratinga8, Carifio24, CarlCarlton, CarlMosel, Carlossuarez46, Carnby, Carson56437, Caryptes, Casper2k3, Catalyst in Society, Catdude, Causa sui, Cause of death, Cautious, Caveman 07, Cedrus-Libani, Celestra, Celyndel, Cenarium, Censusdata, Centrx, Cethegus, Ch'marr, ChKa, Chaddy, Chairboy, Chalst, Chanting Fox, Chaosdruid, CharlesMartel, Charly33, CharonX, Chatanga, Che829, Checkmate987, Chederman69, Cheeseheadxvi, Chicheley, Childzy, Chillum, Chipmunkdavis, Chochopk, Chodorkovskiy, Chokerman88, Choop, Choster, Chowbok, Chris 73, Chris Archer, Chris the speller, Chrisfow, Chrisritch, Christian List, Christinam, Chriswiki, Chrysalis, Chtrede, Chubbles, Chuck Carroll, Chucky21, Church of emacs, Chuunen Baka, Chwyatt, Chzz, Ciacchi, Ciceronl, Ciddler, CieloEstrellado, Cirt, Citizen-of-wiki, City-17, Civil Engineer III, CjDMaX, Cjthellama, Ckatz, Ckorff, Classical 2006, Cliché Online, Cmathio, Cmdrjameson, Coasterlover1994, Coat of Arms, Codycod10, Coffeinfreak, Cogiati, Cogito ergo sumo, Cognate247, Cokes360, Colchicum, ColdWind, Colonies Chris, CommonsDelinker, ConWilBur, Conor's Whipped, Conorbrady.ie, Conorobradaigh, Conte di Cavour, Conti, Contributor777, Conversion script, Cooksey, Coolcole93, Cordless Larry, Coredesat, Cosmic Latte, Crabbit, Craigzomack, Crazycomputers, Crazycoolalex, CrimsonBlue, Crisskriss, Crissov, Crito2161, CrnaGora, Crossk, Crowbar1234, Crusadeonilliteracy, Cruzian, Cryptic, Crystallina, Crzrussian, Cs-wolves, Csörföly D, Cukiger, Culveyhouse, Curps, Curtisvic, Cvieg, Cyde, Cyfal, Cyopardi, Cyrius, D0rkypn0y1029, D34d10gic, D6, DARTH SIDIOUS 2, DCGeist, DGJM, DJ Clayworth, DLiebisch, DMacks, DO'Neil, DTGHYUKLPOQWMNB, DTOx, DVD R W, DaL33T, DaQuirin, Dakhart, Dalmatian Mommy, Damian Radu, Dan D. Ric, Dan Guan, Dan100, DanMS, Dana boomer, Danaman5, Danc, DancingPenguin, Daniel5127, DanielCD, Danielt, Danny, Dantheman531, Dapete, Daqui, Darealclub, DarkFalls, Darkieboy236, Dasani, Dave031970, Dave420, Daveb, Davesmall, David J. Rogers, David Kernow, David Levy, David Liuzzo, David N. Hermann, David.Mestel, Davidfollsom, Davidpdx, Dbachmann, Dca5347, Dcandeto, Deflyer, Deadblob93, Deanos, Debate789, Debresser, Dedeler31, Deejaye6, Deflective, DefunKt, Deiz, Delldot, Deltabeignet, Demiurge, Democryt, Demomoke, Dendodge, DennisDaniels, DennisGigio, Denniss, Deposuit, Deppp, Der Eberswalder, Der Statistiker, Der Wolf im Wald, DerHexer, Deszcz, Deutsche, DevastatorIIC, Devatipan, Deviathan, Deville, Dewritech, Dfrg.msc, Dgg32, Dhp1080, DickLukensTheSecond, Dierk, Dinner123, DirectorR, Directorstratton, DirkvdM, Dirtyhose, DisappointedHorse, Discospinster, Dispenser, DivineIntervention, Dizzyizzy, Djkernen, Djmutex, Dkusic1, Dlugopis, DocendoDiscimus, Doco, Doctor Allen, Docu, Dogface, Dojarca, Donnamaria, Donnerstag, DopefishJustin, Dorftrottel, Doric Loon, Doubleodd, Doug Johnson, Dowj, Downing Street, Dpr, Dr. Crisp, Dr.alf, DrBob, DrKiernan, Dragoburaggo, Dragoon009, Drewbean, Drewrau, Drgals, Driftwood87, Drmies, Drunkenmonkey, Dub4u, Dude8679, Dukeofomnium, Duncharris, Durin, DutchmanJoy, Dysmorodrepanis, Dzenanz, E Pluribus Anthony, E-boy, E235, EBB, ESkog, Easter Monkey, Eatyourgreens, Eco84, Eddospants, Eden Tate, Edgar181, Editor at Large, Editor18, Edwy, EffK, Ehrenkater, Eirik.lingas, Eixo, Ejgreen77, Ekki01, Eklipse, El C, Electionworld, Elekhh, Eleuther, Elf-friend, Elfguy, Elizabeth A, Elliskev, Ellmist, Elmondo21st, Elockid, Eloquence, Elvis, Emperorbma, EncycloPetey, Endofskull, Endurance, EnemyOfTheState, Energyfreezer, Engleaugen6666, Englishnerd, Enlil Ninlil, Enviroboy, EoGuy, Eon, Er Komandante, Eraliswelsh, Erauch, Eric Yurken, EricS, Ericoides, Erlando, ErnieBern, Erstats, Erudra, Esanchez7587, Escape Artist Swyer, EscapingLife, Eschbaumer, Esimal, Etams, Eu.stefan, Euchiasmus, Eugen Simion 14, Eugene van der Pijll, Everyking, Evice, Evil Monkey, F2020, FAThomssen, FF2010, Fabartus, Factuarius, Fahima07, Fainites, Fair Deal, Fallschirmjäger, Fama21, FamilyGuy1998, Familyrits, Farooqmoazzam, Fastestdogever, FatM1ke, Favonian, Fawhad, Feldfrei, Feldmarschall, Felipito1.966, Felixboy, Ferrija1, Ferro Carlotta Monzi Brak, Feuerspiegel,

Fheidener, Fidelio72, Fiet Nam, Filelakeshoe, Finalius, Finlay, FinnishDriver, Firetrap9254, Firsfron, Fisss, Fkbreitl, Flammingo, Flappychappy, Flatterworld, Flauto Dolce, Florentino floro, Florian Blaschke, Flowanda, Flowerpotman, Fnfd, Fonzy, Fooloomanzoo, ForOurXylophones, FordPrkt, Forstnera, Fossa, Fotaun, Four-ArmedRevenge, Fourtimesfour, Franciswiki, Francs2000, Frank Anchor, Freak in the bunnysuit, Freakincoursework, Freakofnurture, Fred Stober, Freddo207, FreplySpang, Friedrichshainer, FritzK0310, Frymaster, Fschoenm, Fsotrain09, Fudge43, Fudoreaper, Fujishima, Fulcher, Funkelnagelneu, Funnyhat, Furrykef, Future Perfect at Sunrise, Fuzheado, Fuzzy510, Fvincent, Fvw, Fwappler, Fyodor, G-Man, G. Campbell, GBWallenstein, GHe, GLGerman, Gabbe, Gabithefirst, Gabriel-Royce, Gadfium, Gaius Cornelius, Gaius Octavius Princeps, Galati, Galoubet, Gamer007, Garryq, Gausie jr, Gavia immer, Gavin Moodie, Gaz, Gekritzl, GenQuest, Geneb1955, Geni, Gennarous, Geo-Loge, Georg Hurtig, GeorgeTSLC, GeraldH, Gerhard51, German Food, German name, GermanJoe, Germanclub, Germanuser, Gert7, Gertjan R., Gertyness, Getker, Gfoley4, Gggh, Ggoere, Ghepeu, GiZiBoNG, Gidonb, Gilbertfein, Gilgamesh, Gilliam, Gilo1969, Gimmetrow, Giovanni76, Girdi, Giving is good, as long as you're getting, Gjp23, Glen, Glenn, Gmaxwell, Gmosoti, GnuDoyng, Godefroy, GoetheFromm, Gogafax, Gogo Dodo, Golbez, Good Olfactory, GoodSirJava, Goodoldpolonius2, Googl, Googleguy1234, Gotrl, Gr8opinionater, Grace Note, Gracenotes, Grade7, GraemeL, Grafen, Grafikm fr, Graham87, Gran2, Graufbold, Greatal386, Greatgreenwhale, GreekNurse, Green Giant, Green Politics, Green caterpillar, GreenUpGreenOut, GregAsche, Grendelkhan, Greyhood, Griffinofwales, GrimmC, Grinman, Grmpf, Gronau, Grstain, Grunt, Gryffindor, Gsklee, Gtg204y, Guanaco, Gugganij, Guidedbyalan, Guideended1, Guilherme Paula, Guliolopez, Gunnar.Kaestle, Guoguo12, Gurch, Gurchzilla, Guy M, Gwernol, Gwyon, Gzornenplatz, H-stt, H0ttamale2, H1523702, HIDECCHI001, HJ Mitchell, HVH, Hadal, Hadseys, HagSibylle, Haham hanuka, Hairy Dude, Hajduk, Hajor, Half price, HalfShadow, Halibutt, Hall Monitor, HamboLSSAH, Hannil, HappyCamper, Harald Meier, Harmil, Harryboyles, Hasan Dalbudak, Hassocks5489, Hatt, Haukurth, Hawkestone, Hayden120, Haymaker, Hazzl, Hdante, Heah, Heimdal, Heimstern, Hektor, Helsinkicommission, Helwerda, Hemanshu, Henry Flower, HerkusMonte, Heron, Herostratus, Herr Kent, Hesus321, Hexil blitz, Hhielscher, Highfields, Highpoems123, Hishii, Historian19, Hmains, Hmwith, Hoc30, Holonboy, Holy Oil And Water, Holynightfever, Homagetocatalonia, Homerjay, Honette, Hoo man, HoodedMan, Hornymanatee, Horst-schlaemma, Househelps, Hqb, Hugo999, HungarianWoman, Husamelodeh, Husond, Hydrargyrum, Hydraton31, I already forgot, IIIraute, IJA, IJzeren Jan, Ian Pitchford, Ibn Battuta, IceArchon, Icecradle, Ich, Ich-AG, Idaltu, Ideletealot2, Ief, Igo4U, Iksisy, Ilehmann, Ilgom, Illythr, Iluvcapra, Ilya1166, Imadcow1, Imaglang, Imlepid, Imperator, ImperatorExercitus, Imperium Europeum, Improv, Imtrogdor, Imwivstuipid, InShaneee, Indon, Infamouskitty, Information-Line, Inge-Lyubov, Ingebjorg, Ingolfson, Instantnood, Int21h, Inwind, Iol, Iowausa, Ipsofactoid, Irdepesca572, Iridescent, Irishguy, Iritebs222, IronGargoyle, Irrypride, IsarSteve, Istabraq, It Is Me Here, Italiano111, Ivirivi00, Ixfd64, Izehar, Izzedine, J Di, J Milburn, J. 'mach' wust, J.K Rowling, J.delanoy, J00tel, JAU123, JCAla, JFT96, JForget, JHMM13, JLogan, JM.Beaubourg, JMS123, JNW, JTBX, Jacek Kendysz, Jack Silver1000, Jackobyte, Jacques2, Jadger, Jaellee, Jaggsan, Jajafe, Jakobburkhardt, James086, James297, Jamesooders, Jamyskis, Jan Hofmann, Janneman, Jansch, Jared Preston, Jarjarbinks10, Jarkeld, Jasman, Jatinbhatt blap, Jatkins, Java13690, Java7837, Jaxl, Jaxsonjo, Jayden54, Jayjg, Jazz+, Jbattersby, Jbou, Jcornelius, Jcrook1987, Jcw69, JdH, JdeJ, JeLuF, Jean.artegui, Jeff G., Jeff3000, Jeffrey O. Gustafson, Jekki, Jeltz, Jensboot, JeremyA, Jeronimo, Jerry67, Jessejamesandloop, Jessica Gordon, Jesushaces, Jesusjonez, Jez, Jfruh, Jhendin, Jiang, Jim bob bryar, Jimiloughlin, Jimp, Jjhcap99, Jjk82, Jklin, Jknabe, Jmcc150, Jmh, JoB614, Joachim Weckermann, JoanneB, JodyB, Joe Canuck, Johan Magnus, Johann Wolfgang, Johannes H.-Larsen, Johanneswilm, John, John D., John K, John R Murray, John Riemann Soong, John Smith M.D., Ph.D, John Smith's, John-vogel, John254, Johnny Undead, Johnpro, Jomika, JonGwynne, JonHarder, JonRoma, Jonah474747, Jonathan Groß, Jondel, Jondumitru, Jonesy, Jonophan, Jooler, Jor, Jose77, Joseph Solis in Australia, Joshua, Jossi, Joséwell, Joy, Jpeob, Jpgordon, Jr724, Jrcrin001, Jrleighton, Jrtsstfjsfkyjsrt6istk7stkyskjytrsk, Jsbhavsar, Jsc83, JuJube, Jujutacular, Julius1990, Jumbuck, Jumpytoo, Junes, Junkietom08, Jusdafax, Jusjih, JustPhil, Justin Eiler, Justin Piga, Justinc, Justinhful, Jwestbrook, Jwordsmith, Jwy, Jyusin, Jza84, KFP, KG200, KJohansson, KMileSupernova, Kacen, Kadillia, Kafziel, KagamiNoMiko, Kahusi, Kaihsu, Kaiser Torikka, Kaiserjd, Kajzderski, Kakofonous, Kalaha, Kamyra, Kantianer, Karam.Anthony.K, Karl, Karl-Henner, KarlFrei, KarlMathiessen, Katherz, KathrynLybarger, Katzensprung, Kbdank71, Kchishol1970, Keegan, Keesiewonder, Keichwa, Keilana, Keith D, Keizuko, Kejaja, Kelly Martin, Kelnor, Kermecke, Kevin, KevinCarmody, KevinOB, KevinTR, Kevs, Kevtrice, Kewp, Kewpid, Khanada, Khoikhoi, Khukri, Killdevil, Killer.k09, KillerChihuahua, Kim-Zhang-Hong, King of Hearts, Kiwifruit89, Kkm010, Klaus Bells, Klausness, Kleuske, Klosterdev, Kman543210, Kman665, Knarf-bz, Knipptang, KnowledgeOfSelf, Knowledgebycoop, Koavf, Komet, Kompakt, Konstantin, Koro lord, Kosebamse, Kotniski, Koyaanis Qatsi, Kozuch, Kr1st1deejay97, KrakatoaKatie, Krankman, Kransky, Krellis, Kribbeh, Kross, Ksenon, Kt66, Kungfuadam, KuroiShiroi, Kuru, Kusma, Kwamikagami, Kwesiidun91, Kwiki, Kyle Benge, Kyle4950, Kyrin, Körnerbrötchen, Kızıl Şaman, L'Etat c'est moi, L1A1 FAL, L95slovenia, LARS, LIU, LKDD, LLLex, LSLM, LaMenta3, Laca, Lacrimosus, LadnaV, Lafuzion, Lakshmix, Lan360, Lancevortex, Landroo, Langston, Larko3344, Larry Sanger, Larry_Sanger, Lars T., Lars Washington, LarsJanZeeuwRules, Laugreve, Laurens-af, Laurentleap, Lawl1232323, Le Fou, Leafyplant, Leandrod, Lear 21, Learngermanwords, Lectonar, Lee J Haywood, Lee S. Svoboda, Leewilson123, Leibniz, Lemsi, Lemsus, Leo Bauer, LeoO3, Lethe, Leuko, Levalley, Levdave7381, Levineps, Levonscott, Lexicon, Lfh, Lhademmor, Lhasaq, Liamdaly620, Liberatus, Lietzow, Liface, LightSpectra, Lightdarkness, Lightmouse, Ligulem, LilHelpa, Lilac Soul, Lilshimon, Lilybaeum, Linuxbeak, Lir, ListedRenegade, LiterallySimon, Lithpiperpilot, LittleDan, LittleOldMe old, LittleRoughRhinestone, LlywelynII, Llywrch, Lobsterantlers, Locke Cole, Lockesdonkey, Logan, Lolhilol, Lolitsjoel16, Lolol123, LonelyMarble, Lonerville, Longbow4u, Longhair, Lord Vader, LordofHavoc, Lordvaluemart, Loren36, Loshaisamazing, Lotje, Louis3ham, Lowellian, Lrobb95, Lstipakov, Lucas323, Lucyin, Ludger1961, Ludjer, Lugnad, LuigiManiac, Luis rib, Luisztdt, Lukatheist, Luna Santin, Lunacy.Fringe, Lunar2, Lunarboy, Lupo, Lurker, Lyfeizshrt, Lysy, MECU, MER-C, MIKElovesgermanchicks, MJCdetroit, MKers06, MKoltnow, MOMBO BAGGINS, MONGO, MPS, MPerel, MTWEmperor, MaGioZal, Macintosh User, Mackensen, Mackeriv, Mactruth, Madman, Maelnuneb, Magicjigpipe, Magister Mathematicae, Magnus Manske, Magumbaboy, Mailer diablo, Malarious, Malbi, Malcolm Farmer, Male1979, Malerin, Malfidus, Malo, Man vyi, Mandavi, Manecke, Maneus4, Mangoe, Manop, Manpyo, MapsMan, Marcika, Marco polo, Marcus Cyron, Marcworca, Marek69, MarekF, Maria Stella, Marielleh, Mark, Mark.P.Bartlett, MarkThomas, Markus Schmaus, Marky-Son, Marlight, Marnanel, Marriex, MarsRover, Martg76, Martin.schoenberg, Martinp23, Martinwilke1980, Maru4u, Marvan, Marysunshine, Marzoregulus@yahoo.co.ph, Marzter6, Masken, Masonshowalter, Massjit, Master Jay, MasterLycidas, Materialscientist, MathFacts, Mathew123, MathiasFox, Matilda, Matros, Matstone99, Matt 989, Matt Crypto, Matt512, Mattbaileyisneat, Matthead, Matthew Fennell, Matthewjw 16, Matthäus Wander, Matticus78, Mattingly23, MatttK, Mauriziotani, Mav, Maverick starstrider, Mawa, MaxBech1975, Maximus Rex, Maxschmelling, Maymose, Mayonnaise, Mazzelino, McDogm, McGeddon, Mcewan, Mclay1, Mcmoor, Mcurmamma, Me Mighty Leper, Meelar, MegA, Meh31, Mehmeda, MeinGoshen, Melaen, Mellum, MementoVivere, Menameischarle, Menchi, Mentifisto, Merbabu, Merchbow, Mercury, Merlincooper, Mertan, Mervyn, Mesgul82, Metaljaws, Mfa fariz, Mgiganteus1, Mgoodyear, MiLo28, Miaers, Mic, Mich112358, Michael Hardy, Michael Zimmermann, Michael riber jorgensen, MichaelDiederich, Michaelas10, Michaelmcneil, Michaeltobbylee, Michbich, Michealkelly, Michel rockstar, Mickey gfss2007, Microsomen, Middleeast, Midnightblueowl, Mifter, Mightymights, Miguel Andrade, Miguel.mateo, Mike Rosoft, Mike.heckmann, MikeAtari, Mikeifc123, MillaSukel, Milton Stanley, Mipago, Miroslawa, MisfitToys, Miskin, Missspencer06, Mister C4, MisterSheik, Misza13, Mittermeier, Mixcoatl, Miyokan, Mjpieters, Mmunger, Mnemeson, Mnmazur, Mo-Al, Mo0, Modelun88, Moe Epsilon, Moehre, Mollemy, Molobo, Moncrief, Monkeyknife, Monomachus, Moominjosh, Moonunit02, Mooseberry, Moriori, Morse thor, Morwen, Moshe Constantine Hassan Al-Silverburg, Mouvement, Movementarian, Moxdude, Moxy, Mqduck, Mr Adequate, Mr MillA, Mr Stephen, Mr Woman, Mr. Know-It-All, Mr. bwn, MrCanadian1, MrFish, MrPMonday, MrRandomPerson, MrTranscript, Mramz88, Mrfathead, Mrjeff, Mschel, Mshecket, Mtaylor848, Mtl 16, Mtmoore321, Mucky Duck, Mufka, Munci, Mus Musculus, Muskydusky, Mxcatania, Mxn, My76Strat, Myanw, Myscrnnm, N-edits, N5iln, NB-NB, NYSTEX1992, Naive cynic, Nakon, NameIsRon, Narya, Naryathegreat, Nascar1996, Natalie Erin, Naverno, NawlinWiki, Nbatra, Nebauer, Nectarflowed, Nellov5, Neozoon, Nero the second, NeroN BG, NetherlandishYankee, Netknowle, Netsnipe, Neurowin, Neutrality, Nevfennas, NewYork483, Newkai, Newnoise, Nicander, Nichalp, Nick San, Nicketje666, Nickinny, Nickispeaki, Nickj, Nicklovesgold, Nicksukh, Nico, Nicofs, Nicolas1995, Nieding, NigelR, Night w, Nihil novi, Nika, Nikkimaria, Niko226, Nikola Smolenski, Nikostar, Nimbusania, Nirvana888, Nishkid64, Nitya Dharma, Nivix, Nixdorf, Njecoolguy90, Nlnnet, Nmpenguin, NoGringo, NoPuzzleStranger, Nobleeagle, Nohomers48, Nolelover, Nonpawit, Noon, Noozgroop, Nopetro, Northernmonkey, Nothlit, Notuncurious, Nouse4aname, Npovshark, Nricardo, Nscheffey, Ntwarrior, NuclearVacuum, NuclearWarfare, Nufy8, Numbo3, Nunquam Dormio, Nv8200p, Nzpcmad, OOODDD, OSX, Obarskyr, Octane, Od Mishehu, OettingerCroat, Ohconfucius, Ohnoitsjamie, Okichubiness, Okinawadude, Olaf Eriksson, Old Guard, Oldnag85, Olegwiki, Olessi, Olgaru, OliFer, OliverH, OlliffeObscurity, Olorin28, Olowek, Oman66, Omicronpersei8, Onco p53, OneGuy, Oneearth, Onlineinteraction, Oo64eva, Oosoom, Opc123, Opelio, Opus88888, Orangerider, OrbitOne, Oreo Priest, Ori.livneh, Ornithologician, Orz, Ospalh, Ot, Ottawa4ever, Otto ter Haar, OwenX, P-Schmidt22, P.juka, PFHLai, PS2pcGAMER, PSWG1920, Paddyez, Paine Ellsworth, Paintball12345, Pakhomovru, Palica, Pan Dan, Panoramastitcher, Pantherskin, Paradocks2, Paranoid, Pascal.Tesson, Pathless, Patka, Patrick, Patstuart, Patyk, Paul Benjamin Austin, Paul Erik, PaulBommel, Pavlo Chemist, Pawel z Niepolomic, Pax:Vobiscum, Pearle, Pecher, Penmachine, Pensionero, Pepsidrinka, Per Abrahamsen, Peter, Peter Horn, Peter Isotalo, Peter2005, PeterCS, Peterale4444, Pethr, Petree13, Petri Krohn, Petrux, Pewwer42, Pgk, Phaedriel, Phantomsteve, Pharaoh of the Wizards, Phatom87, Phenz, Phil Bastian, Phil Boswell, PhilKnight, Philip Baird Shearer, Philip Cross, Philip ea, Phillip J, Philophobia, Philter137, Phoenix of9, Picapica, Picaroon, Piccadilly, Piggyspizza, Pilotguy, Piotrus, Piratelooksat30some2001, Pirawik, Pjacobi, Platane, Plocho22, Plumber, Poarps, Pofka, Pokrajac, Polaron, Polylepsis, Pommerland, Popadopolis, Porqin, Porsche997SBS, Postcard Cathy, Postdlf, Postmann Michael, Prattlement, Preslethe, Primate, PrimeCupEevee, Pristino, ProhibitOnions, Proteus, Prumpf, Pschemp, Pseudo-Richard, PseudoSudo, Psg116, Pslide, Psy guy, Puchiko, Punctilius, Purple Sheep, Puttputttime, PuzzletChung, Pyroblondie, Qelp, Qero, Quakenul, QuartierLatin1968, QueenStupid, Queson, Quibik, Quizimodo, Quizimodo (usurped), Qweedc, Qwerty Binary, R Lowry, R'n'B, R-41, R9tgokunks, RG2, RHB, RM21, RMHED, RPlunk, RSido, Rada, Radagast, Radu Gherasim, RaffiKojian, RagingR2, Railer 353, Ralphy, Ran, Randomblue, RandySavageFTW, RapaNui75, Rarelibra, Raul14, Raul654, Raven4x4x, Ray Van De Walker, Raymond arritt, Rbz321, Rd232, Reboot, Reconsider the static, Recury, Red Winged Duck, RedWolf, Reddi, Redking7, Rednaxela, Reedy, Reimanova, Reindra, Reinis, Reisio, Rejedef, Renata, ResearchRave, Resilienzi, Retired username, Retireduser1111, Rettetast, Revolving Bugbear, Rex Germanus, RexNL, Reywas92, Rflesher, Rhys fry, Riana, Ribbet32, Ricbep, Rich Farmbrough, Richard D. LeCour, Richardcavell, Richarddd, Richardfabi, Rick Block, RickK, Ricky@36, Rickyrab, Rilekou, Rimshot, Rje, Rjensen, Rjwilmsi, Rl, Rmky87, Rmt2m, Rnelson, RoMex, RoadTrain, Rob 06-07, Rob.derosa, Robdurbar, Robendo, Robert1947, Robert25, RobertG, Robertsteadman, Robinstocks, Robomaeyhem, Robzz, Rocastelo, RockMFR, Rocket000, Rockfang, Rockstar4000, Rod57, Rogerd, Rokbas, Romankawe, Romanm, Ronhjones, Rosenzweig, Routerrandy4, Roxi2, Roy da Vinci, RoyBoy, Royboycrashfan, Rracecarr, Rreagan007, Rtc, RuffRyderMohan, Ruhrjung, Rururuu, RussBlau, Ruud64, RxS, RyanGerbil10, Ryulong, SFC9394, SKC, SNIyer12, Sacxpert, Saforrest, Sagaciousuk, Said 13, Saimdusan, Saintswithin, Salion, SallyForth123, Salmnj, Salsa Shark, Salsb, Sam Korn, Sam Spade, Sam Vimes, Sam-m, Sam01257462071, SamEV, Samantha555, Samashton, Samgardner94, Samir, Samrolken, Samsara, Samuel, Samuelsen, Sandersonjoe221, Sandman, Sandpiper, Sandwriter, SandyGeorgia, Sango123, Sannse, Sarcelles, Sardanaphalus, Satori, Savidan, Saxxtreme, Sca, Scartboy, Sceptre, Schaeferchr11, Schaengel89, Scheibenzahl, Schlumpff, Schneelocke, Schultz.Ryan, Schumacherfan, SchuminWeb, Schwatdy, Schzmo, Scifiintel, Scipius, Sciptor, Sciurinæ, Scohoust, Scott777, Scovetta, Scrollriter, Scryer 360, Scythian1, Seaman, Sean William, Seb az86556, Sebesta, Secret (renamed), Section6, SeeSchloss, Seksen iki yüz kırk beş, Selinger, Semifriend, Semprini, Separa, Serafin, SergioGeorgini, Serpicotv, Serte, Sevela.p, Sfahey, Sgeureka, Sgrayban, Sgt Simpson, Shadiac, Shadow007, Shadow1, Shadowjams, Shamir1, Shanel, Shanes, Shannon bohle, Shastrix, Shenhemu, Sheogorath, Sherollin, Sherool, Shimgray, Shivan, Shlomke, Shoaler, Short Verses, Shredder2012, Shreshth91, Sicilianmandolin, SigPig, Signsolid, Simetrical, SimonP, Simonides, Sintaku, Sir Isaac, Sir Richardson, Sirius85, Sjacinth, Sjakkalle, Sketchmoose, Skinnyweed, Skizzik, Sky, Sky Diva, Skysmith, Skywolf, Skäpperöd, SlamDiego, Sleigh, Sleptrip, Slomox, Slowking Man, SluggoOne, Smaalouf, Smith2006, SmthManly, Smug Irony, Smyth, Snack417, Snagari, Snazzcen, Snickers45, Snieg, SnowFire, Snowolf, Snoyes, SoCalSuperEagle, SoLando, SoWhy, Softjuice, Solicitr, SolidEdgeGuRu, Solipsist, SomeHuman, Something Wicked, Somethingoranother, Son of Al, Soriamic, Sortior, Sousclef, Soxwon, Sp33dyphil, SpNeo, Space Cadet, SpamBilly, Specialplant, Spitfire, Spitfire19, Spitzl, Splodgeness, Spongefrog, SpookyMulder, Springeragh, Springwalk, Spyro02, Squalie21, SqueakBox, Squirrelofdoom5, Srich32977, Srndp8e, Ssault, Ssolbergj, Staatenloser, Stallions2010, Stangaa, Steel, Stefan Udrea, Stemonitis, Stephan Schulz, Stephen G. Brown, Stephenb, Stephreef, Stern, Stestagg, Stetay, Stevemangold, Stevencool104, Stevepeterson, Stevouk, Stickee, Stikshift, StoneColdCrazy, StoneProphet, Stonnman, Stor stark7, Stormscape, Str1977, Stratd00d, StuartBrady, Stuartrocks@gmail.com, Student7, Stuffit, Subseven26, Subtropical-man,

Suisui, Sullengrrl, SunDog, Sundar, Sundar1, Sunray, Supadawg, SuperHamster, Supertask, Supreme Deliciousness, Surlyduff50, Suruena, Suryoye, Sushi Leone, Sux2ussj, SvenskaJohannes, Swatjester, Swedish fusilier, Swid, Synthe, Syrthiss, Syslord2k, Szajci, Sümpf, T2Green, TFCforever, TGC55, THEunique, THX-1138, TSO1D, TShilo12, TUF-KAT, Ta bu shi da yu, Tacen28, Talorrob, Tangotango, Tapir Terrific, Targeman, Tariqabjotu, Tarret, Tartessos75, Tasc, TaurusMount, Taw, Tawker, Taxman, Tb, Tbacker, Tbhotch, Tcooke, Teccen, Tecomaster345, TedE, Teddycom, Tedp, Teh tennisman, Template namespace initialisation script, TenOfAllTrades, Terence, Teryx, Teuton9, Teutsch, TexasAndroid, Texture, Tezero, Tfine80, Tfjt, Thabet202, Thaejas, Thayts, The Anome, The Catfish, The Clinkynator, The Emirr, The Mysterious El Willstro, The Phoenix, The Thing That Should Not Be, The Tom, The Transhumanist, The Utahraptor, The demiurge, The dog2, The misha, The monkeyhate, The russian leader, The undertow, TheFox2, TheK, TheKMan, TheLewisRepublic, ThePointblank, TheTrojanHought, Theentireending, Thefullback, Thegoodson, Themanwithoutapast, Theonlyedge, Therealdrag0, Therequiembellishere, Thewallowmaker, Thewikipedian, Thing that goes on feet, Thingg, Think outside the box, Thomas Ganter, Thomas Ludwig, ThomasK, Thore, Threepwood89, Thricecube, Thughes1774, Thumperward, Thunk, Thuresson, Tiddle 999, Tiddly Tom, Tiki99, Tillwe, Tim buckley, TimBRoy, TimBentley, Timeineurope, Timmothias, Timwi, Tintenfischlein, Titoxd, Tiwonk, Tjl1128, Tkgd2007, Tkinias, Tkynerd, Tmasuda, Tobby72, Tobyw87, Tocino, Toffel07, Tokikake, Tokiohotelover, Tolaagboola, Tom harrison, TomR, TomaK, Tomeasy, Tomfreelance, Tomi, Tommy2010, Tommycw1, Tomorrow Never Knows, Tompayne1976, Tone, Tony Sidaway, Toothpull, Toussaint, TownDown, Tpbradbury, Tra, Trasman, Travelbird, Traveletti, TrbleClef, TreXiaNos, Tree Biting Conspiracy, Treisijs, Trewin, Treyt021, Trialsanderrors, Trickstar, Triggerpull2, Trilobite, Trinitrix, Troy 07, Trueshow111, Tsujigiri, Ttownfeen, Tulkolahten, Tunheim, Tuomas, TurabianNights, Twilight Princess 64, Twoshirou, Tyduro, Tyrol5, Tzartzam, Tzustrategy, U.S.A.U.S.A.U.S.A., U.Steele, UKWiki, UNDGRND, Uannis, Ubermonkey, Ufwuct, Ugur Basak, UkPaolo, Ulf Heinsohn, Ulitzde, Ulmanor, Ulritz, Ultimate Destiny, UncleAdolf, Underlying lk, Unify, Universe=atom, Unreal595, Unused0029, Unyoyega, Urban XII, Uriber, Ursutraide, User F203, User27091, UserDoe, Usergreatpower, Utcursch, Ute in DC, Uurtamo, V6g3h7, Valentinian, Valentino, Valleyhome, Van helsing, Vancouverguy, Vanderdecken, Vaniba12, Vanis, Vardion, Varlaam, Vary, Vdjj1960, Veit, Veledan, Velho, Ventur, Verdadero, VeryVerily, Vexorg, Vextration, Victor Hugo, Viewfinder, Vilallonga, Vincentnufcr1, Violetriga, Vir, Virgule82, VitaleBaby, Vkem, VolatileChemical, Vondort, Vortexx, Vranak, Vrghs jacob, Vsmith, Vulcant13, Vzbs34, WHEELER, WTGDMan1986, WTucker, Wackojacko120, Wafulz, Walkerma, Wangi, Waprap, Warfwar3, Warhead66, Warofdreams, Watbjbjbjk, Wayward, We hope, Weareqpr, WegianWarrior, Weissmann, Welle, Welshleprechaun, WereWolf, Wernergerman, Westb3ach, Wester, Whaatt, WhisperToMe, Whitey115, Who, Whouk, Wiatr, Wik, Wiki alf, WikiBrown, Wikiain, Wikiborg, Wikifriendawesome, Wikipedia-WillYouMarryMe?, Wikipelli, Wikiperuvian, Wikist, Wikithor, Will Beback Auto, Willbur, Williamborg, Williame3, Williamwade, Willyhunter123, Wimt, Windsamurai, Winhunter, Wireless friend, WishingForever, Wissahickon Creek, Wk muriithi, Wknight94, Wliiam·Shakespeare, Wlmg, Wolf530, Woohookitty, Wootini, Worm That Turned, Wow rules the world, WriterHound, WritersCramp, Wulf Isebrand, XLerate, XZise, Xaosflux, Xenus, Xeph, Xezbeth, Xigan, Xtopher, YAG490, Yahel Guhan, Yamamoto Ichiro, Yandman, YankeeDoodle14, YellowMonkey, Yngvadottir, Yoda1893, Yoghurt, Yoki11, Yorkshirian, Youknowyouknow, Youmuppet, Youngamerican, YourEyesOnly, Yousou, Youssefsan, Yvesdesprit, Yzmo, Z.E.R.O., Zacefron52, Zack Holly Venturi, Zahnradzacken, Zaindy87, ZakuSage, Zaphodia, Zarbi1, Zarniwoot, Zeldafan3242, Zeno Gantner, Zesel, Zhou Yu, Zia demion, Ziusudra, Zmaj, Zondor, Zoney, Zpatel93, Zscout370, Zsinj, Zundark, Zzuuzz, Zzyzx11, £20, °, Äpple, Ævar Arnfjörð Bjarmason, Île flottante, Özdemir, ΚΕΚΡΩΨ, Методије, טראהגייר ירעל, نولیسپا, نيلرب رسن, €pa, 4458 anonymous edits

Image Sources, Licenses and Contributors

Image:Wittenberge Panorama2.jpg *Source*: http://en.wikipedia.org/w/index.php?title=File:Wittenberge_Panorama2.jpg *License*: unknown *Contributors*: User:Niteshift

File:Wappen_Wittenberge.png *Source*: http://en.wikipedia.org/w/index.php?title=File:Wappen_Wittenberge.png *License*: unknown *Contributors*: Knorrepoes, Magul, Moros, Niteshift, Ollemarkeagle, Rauenstein

file:Germany location map.svg *Source*: http://en.wikipedia.org/w/index.php?title=File:Germany_location_map.svg *License*: unknown *Contributors*: User:NordNordWest

File:Red_pog.svg *Source*: http://en.wikipedia.org/w/index.php?title=File:Red_pog.svg *License*: unknown *Contributors*: Anomie

Image:Wittenberge Rathaus.jpg *Source*: http://en.wikipedia.org/w/index.php?title=File:Wittenberge_Rathaus.jpg *License*: unknown *Contributors*: Original uploader was Cepheiden at de.wikipedia

File:Westerrönfeld Wappen.png *Source*: http://en.wikipedia.org/w/index.php?title=File:Westerrönfeld_Wappen.png *License*: unknown *Contributors*: User:AndreasPraefcke

Image:Administrative Gliederung Deutschlands.png *Source*: http://en.wikipedia.org/w/index.php?title=File:Administrative_Gliederung_Deutschlands.png *License*: unknown *Contributors*: User:David Liuzzo

File:Administrative Districts of Germany 1981-2008.svg *Source*: http://en.wikipedia.org/w/index.php?title=File:Administrative_Districts_of_Germany_1981-2008.svg *License*: unknown *Contributors*: Bennet Schulte

File:Schleswig-Holstein_RD.svg *Source*: http://en.wikipedia.org/w/index.php?title=File:Schleswig-Holstein_RD.svg *License*: unknown *Contributors*: User:Hagar66

File:Flag of Germany.svg *Source*: http://en.wikipedia.org/w/index.php?title=File:Flag_of_Germany.svg *License*: unknown *Contributors*: Anomie

Image:Wappen Kreis Rendsburg-Eckernfoerde.svg *Source*: http://en.wikipedia.org/w/index.php?title=File:Wappen_Kreis_Rendsburg-Eckernfoerde.svg *License*: unknown *Contributors*: User:MorbZ

Image:Uebersicht-RD.png *Source*: http://en.wikipedia.org/w/index.php?title=File:Uebersicht-RD.png *License*: unknown *Contributors*: ClausG

File:Flag of Schleswig-Holstein.svg *Source*: http://en.wikipedia.org/w/index.php?title=File:Flag_of_Schleswig-Holstein.svg *License*: unknown *Contributors*: User:Jhs

File:Landeswappen Schleswig-Holstein.png *Source*: http://en.wikipedia.org/w/index.php?title=File:Landeswappen_Schleswig-Holstein.png *License*: unknown *Contributors*: Schleswig-Holstein

File:Deutschland Lage von Schleswig-Holstein.svg *Source*: http://en.wikipedia.org/w/index.php?title=File:Deutschland_Lage_von_Schleswig-Holstein.svg *License*: unknown *Contributors*: User:David Liuzzo

File:Speaker Icon.svg *Source*: http://en.wikipedia.org/w/index.php?title=File:Speaker_Icon.svg *License*: unknown *Contributors*: Blast, G.Hagedorn, Mobius, 2 anonymous edits

File:Limes.saxoniae.wmt.png *Source*: http://en.wikipedia.org/w/index.php?title=File:Limes.saxoniae.wmt.png *License*: unknown *Contributors*: User:Wmeinhart

File:KielerStadtzentrumLuftaufnahme.jpg *Source*: http://en.wikipedia.org/w/index.php?title=File:KielerStadtzentrumLuftaufnahme.jpg *License*: unknown *Contributors*: User:Drbashir117

File:Lubeck panorama.JPG *Source*: http://en.wikipedia.org/w/index.php?title=File:Lubeck_panorama.JPG *License*: unknown *Contributors*: Original uploader was Innomann at de.wikipedia (Original text : Innomann)

File:Rape-fieldSH.jpg *Source*: http://en.wikipedia.org/w/index.php?title=File:Rape-fieldSH.jpg *License*: unknown *Contributors*: User:Jvdc

File:Sylt beach.jpg *Source*: http://en.wikipedia.org/w/index.php?title=File:Sylt_beach.jpg *License*: unknown *Contributors*: Sean O'Flaherty at en.wikipedia

Image:Abstimmung-schleswig-1920.png *Source*: http://en.wikipedia.org/w/index.php?title=File:Abstimmung-schleswig-1920.png *License*: unknown *Contributors*: Bennet Schulte

File:Schleswig-Holstein.png *Source*: http://en.wikipedia.org/w/index.php?title=File:Schleswig-Holstein.png *License*: unknown *Contributors*: User:Ulamm

File:Schleswig-Holstein Kreise (nummeriert).svg *Source*: http://en.wikipedia.org/w/index.php?title=File:Schleswig-Holstein_Kreise_(nummeriert).svg *License*: unknown *Contributors*: User:David Liuzzo

File:Rote Grütze mit Vanillesoße.JPG *Source*: http://en.wikipedia.org/w/index.php?title=File:Rote_Grütze_mit_Vanillesoße.JPG *License*: unknown *Contributors*: User:Olaf Simons

File:Helgoland Hummerbuden 21955.JPG *Source*: http://en.wikipedia.org/w/index.php?title=File:Helgoland_Hummerbuden_21955.JPG *License*: unknown *Contributors*: User:Josi

File:Coat of Arms of Germany.svg *Source*: http://en.wikipedia.org/w/index.php?title=File:Coat_of_Arms_of_Germany.svg *License*: unknown *Contributors*: Wappenentwurf: Karl-Tobias Schwab (1887–1967), entworfen 1926 diese Datei: Jwnabd

File:EU-Germany.svg *Source*: http://en.wikipedia.org/w/index.php?title=File:EU-Germany.svg *License*: unknown *Contributors*: User:NuclearVacuum

File:Increase2.svg *Source*: http://en.wikipedia.org/w/index.php?title=File:Increase2.svg *License*: unknown *Contributors*: Sarang

Image:Speakerlink.svg *Source*: http://en.wikipedia.org/w/index.php?title=File:Speakerlink.svg *License*: unknown *Contributors*: Woodstone. Original uploader was Woodstone at en.wikipedia

File:Imperium Romanum Germania.png *Source*: http://en.wikipedia.org/w/index.php?title=File:Imperium_Romanum_Germania.png *License*: unknown *Contributors*: Modification · Bearbeitung · Prilaboro: D. Bachmann

File:Martin Luther by Lucas Cranach der Ältere.jpeg *Source*: http://en.wikipedia.org/w/index.php?title=File:Martin_Luther_by_Lucas_Cranach_der_Ältere.jpeg *License*: unknown *Contributors*: 1970gemini, Aavindraa, Avatar, Con2tto, DaTroll, Dr. Shaggeman, FA2010, Interpretix, Krinkle, Miya, Origamiemensch, Pomfuttge, Quadell, Rosenzweig, Sailko, Semnoz, Shakko, Warburg, Wst, Ævar Arnfjörð Bjarmason, 3 anonymous edits

File:Wernerprokla.jpg *Source*: http://en.wikipedia.org/w/index.php?title=File:Wernerprokla.jpg *License*: unknown *Contributors*: 1970gemini, 52 Pickup, Adam Cuerden, AnRo0002, Chrisportelli, DIREKTOR, Jacklee, Jappalang, Kameraad Pjotr, Kresspahl, Leppus, Mattes, Mnmazur, Mogelzahn, Shakko, Túrelio, Vanished user 001, Wuselig, Ælfgar, 7 anonymous edits

File:Bundesarchiv Bild 183-S33882, Adolf Hitler retouched.jpg *Source*: http://en.wikipedia.org/w/index.php?title=File:Bundesarchiv_Bild_183-S33882,_Adolf_Hitler_retouched.jpg *License*: unknown *Contributors*: unknown

File:Potsdamer Platz 1945.jpg *Source*: http://en.wikipedia.org/w/index.php?title=File:Potsdamer_Platz_1945.jpg *License*: unknown *Contributors*: Michael M. Dean (Canada. Dept. of National Defence/Ministère de la défense nationale)

File:Map-Germany-1945.svg *Source*: http://en.wikipedia.org/w/index.php?title=File:Map-Germany-1945.svg *License*: unknown *Contributors*: Original uploader was 52 Pickup at en.wikipedia

File:Thefalloftheberlinwall1989.JPG *Source*: http://en.wikipedia.org/w/index.php?title=File:Thefalloftheberlinwall1989.JPG *License*: unknown *Contributors*: Unknown photographer, Reproduction by Lear 21

File:Deutschland topo.jpg *Source*: http://en.wikipedia.org/w/index.php?title=File:Deutschland_topo.jpg *License*: unknown *Contributors*: User:Botaurus-stellaris

File:Aquila chrysaetos 3 (Martin Mecnarowski).jpg *Source*: http://en.wikipedia.org/w/index.php?title=File:Aquila_chrysaetos_3_(Martin_Mecnarowski).jpg *License*: unknown *Contributors*: User:Trachemys/Martin Mecnarowski

File:Reichstag building Berlin view from west before sunset.jpg *Source*: http://en.wikipedia.org/w/index.php?title=File:Reichstag_building_Berlin_view_from_west_before_sunset.jpg *License*: unknown *Contributors*: User:JürgenMatern

File:2011 Joachim Gauck-2.jpg *Source*: http://en.wikipedia.org/w/index.php?title=File:2011_Joachim_Gauck-2.jpg *License*: unknown *Contributors*: User:Hic et nunc, User:J. Patrick Fischer

Image:Karte Bundesrepublik Deutschland.svg *Source*: http://en.wikipedia.org/w/index.php?title=File:Karte_Bundesrepublik_Deutschland.svg *License*: unknown *Contributors*: User:David Liuzzo

Image:Coat of arms of Lower Saxony.svg *Source*: http://en.wikipedia.org/w/index.php?title=File:Coat_of_arms_of_Lower_Saxony.svg *License*: unknown *Contributors*: Darwinius, Dbenbenn, Fred the Oyster, Frombenny, Jimmy44, Lou.gruber, Ludger1961, Madden, Mattes, Ollemarkeagle, Pitke, Rosenzweig, 2 anonymous edits

Image:Bremen Wappen(Mittel).svg *Source*: http://en.wikipedia.org/w/index.php?title=File:Bremen_Wappen(Mittel).svg *License*: unknown *Contributors*: User:David Liuzzo

Image:Coat of arms of Hamburg.svg *Source*: http://en.wikipedia.org/w/index.php?title=File:Coat_of_arms_of_Hamburg.svg *License*: unknown *Contributors*: Frombenny, GeorgHH, Havang(nl), Jodo, Leit, Madden, Milgesch, Mogelzahn, N3MO, Rosenzweig, Warburg, Überraschungsbilder, 7 anonymous edits

Image:Coat of arms of Mecklenburg-Western Pomerania (great).svg *Source*: http://en.wikipedia.org/w/index.php?title=File:Coat_of_arms_of_Mecklenburg-Western_Pomerania_(great).svg *License*: unknown *Contributors*: Bundesland Mecklenburg-Vorpommern

Image:Wappen Sachsen-Anhalt.svg *Source*: http://en.wikipedia.org/w/index.php?title=File:Wappen_Sachsen-Anhalt.svg *License*: unknown *Contributors*: Land Sachsen-Anhalt

Image:Coat of arms of Saxony.svg *Source*: http://en.wikipedia.org/w/index.php?title=File:Coat_of_arms_of_Saxony.svg *License*: unknown *Contributors*: Freistaat Sachsen; House of Wettin

Image:Brandenburg Wappen.svg *Source*: http://en.wikipedia.org/w/index.php?title=File:Brandenburg_Wappen.svg *License*: unknown *Contributors*: User:David Liuzzo

Image:Coat of arms of Berlin.svg *Source*: http://en.wikipedia.org/w/index.php?title=File:Coat_of_arms_of_Berlin.svg *License*: unknown *Contributors*: Wappenentwurf: diese Datei: Jwnabd

Image:Coat of arms of Thuringia.svg *Source*: http://en.wikipedia.org/w/index.php?title=File:Coat_of_arms_of_Thuringia.svg *License*: unknown *Contributors*: Burts, Frombenny, Ludger1961, MB-one, Madden, Michael Sander, Mogelzahn

Image:Coat of arms of Hesse.svg *Source*: http://en.wikipedia.org/w/index.php?title=File:Coat_of_arms_of_Hesse.svg *License*: unknown *Contributors*: Darwinius, Ewan McTeagle, F. F. Fjodor, Frombenny, Madden, Mogelzahn, Perhelion, Rauenstein, Túrelio, 1 anonymous edits

Image:Coat of arms of North Rhine-Westfalia.svg *Source*: http://en.wikipedia.org/w/index.php?title=File:Coat_of_arms_of_North_Rhine-Westfalia.svg *License*: unknown *Contributors*: Wappenentwurf (1947): Wolfgang Pagenstecher für das Land Nordrhein-Westfalen

Image:Coat of arms of Rhineland-Palatinate.svg *Source*: http://en.wikipedia.org/w/index.php?title=File:Coat_of_arms_of_Rhineland-Palatinate.svg *License*: unknown *Contributors*: Land Rheinland-Pfalz

Image:Bayern Wappen.svg *Source*: http://en.wikipedia.org/w/index.php?title=File:Bayern_Wappen.svg *License*: unknown *Contributors*: User:David Liuzzo

Image:Coat of arms of Baden-Württemberg (lesser).svg *Source*: http://en.wikipedia.org/w/index.php?title=File:Coat_of_arms_of_Baden-Württemberg_(lesser).svg *License*: unknown *Contributors*: AndreasPraefcke, BrightRaven, Darwinius, Ewan McTeagle, Frombenny, Gryffindor, Jed, Madden, Milgesch, Rtc, Savh, Ssch, 2 anonymous edits

Image:Coa de-saarland.svg *Source*: http://en.wikipedia.org/w/index.php?title=File:Coa_de-saarland.svg *License*: unknown *Contributors*: User:Chrkl

Image:Coat of arms of Schleswig-Holstein.svg *Source*: http://en.wikipedia.org/w/index.php?title=File:Coat_of_arms_of_Schleswig-Holstein.svg *License*: unknown *Contributors*: Burts, Drbashir117, Frombenny, Mogelzahn, Rosenzweig, Shizhao, 1 anonymous edits

File:33rdG8Leaders.jpg *Source*: http://en.wikipedia.org/w/index.php?title=File:33rdG8Leaders.jpg *License*: unknown *Contributors*: White House photo by Eric Draper

File:S-Klasse W221.jpg *Source*: http://en.wikipedia.org/w/index.php?title=File:S-Klasse_W221.jpg *License*: unknown *Contributors*: Fadi.

File:Germany Product Export Treemap.jpg *Source*: http://en.wikipedia.org/w/index.php?title=File:Germany_Product_Export_Treemap.jpg *License*: unknown *Contributors*: 1Veertje, Cambalachero, Doubleodd

File:ICE 3M KRM.jpg *Source*: http://en.wikipedia.org/w/index.php?title=File:ICE_3M_KRM.jpg *License*: unknown *Contributors*: User:Sese_Ingolstadt

File:Einstein 1921 by F Schmutzer.jpg *Source*: http://en.wikipedia.org/w/index.php?title=File:Einstein_1921_by_F_Schmutzer.jpg *License*: unknown *Contributors*: Calliopejen1, DIREKTOR, Frank C. Müller, Hemulen, Lobo, Quibik, Raeky, Андрей Романенко

File:Population Pyramid of Germany in 2012.png *Source*: http://en.wikipedia.org/w/index.php?title=File:Population_Pyramid_of_Germany_in_2012.png *License*: unknown *Contributors*: International Futures (IFs)

File:Kdom.jpg *Source*: http://en.wikipedia.org/w/index.php?title=File:Kdom.jpg *License*: unknown *Contributors*: ger1axg

File:Heidelberg Universitätsbibliothek 2003.jpg *Source*: http://en.wikipedia.org/w/index.php?title=File:Heidelberg_Universitätsbibliothek_2003.jpg *License*: unknown *Contributors*: Jan Beckendorf

File:Beethoven.jpg *Source*: http://en.wikipedia.org/w/index.php?title=File:Beethoven.jpg *License*: unknown *Contributors*: ADGE, Aaron1a12, Butko, Codeispoetry, CommonsDelinker, Herbythyme, Hotshot977, Howcheng, Ies, Interpretix, Jappalang, Johney, Jusjih, Leonce49, Mattes, Michael Bednarek, Neukoln, Pixel ;-), Shakko, Sonphan10, Thuresson, אמסיסה123, 4 anonymous edits

File:Grimm.jpg *Source*: http://en.wikipedia.org/w/index.php?title=File:Grimm.jpg *License*: unknown *Contributors*: AndreasPraefcke, Anne-Sophie Ofrim, Bukk, Gabor, Hannah, Kresspahl, Léna, Mattes, Nillerdk, Origamiemensch, Phrood, Reinhard Dietrich, Shakko, Yuval Y, 1 anonymous edits

File:Black Forest gateau.jpg *Source*: http://en.wikipedia.org/w/index.php?title=File:Black_Forest_gateau.jpg *License*: unknown *Contributors*: Mikelo

File:Westfalenstadion von oben.jpg *Source*: http://en.wikipedia.org/w/index.php?title=File:Westfalenstadion_von_oben.jpg *License*: unknown *Contributors*: Druffeler, EugeneZelenko, Fuzzy510, Hornymanatee, Mbdortmund, Odin statt Jesus, RHunscher, STBR

File:Openstreetmap logo.svg *Source*: http://en.wikipedia.org/w/index.php?title=File:Openstreetmap_logo.svg *License*: unknown *Contributors*: OpenStreetMap

GNU Free Documentation License Version 1.2, November 2002 Copyright (C) 2000,2001,2002 Free Software Foundation, Inc. 59 Temple Place, Suite 330, Boston, MA 02111-1307 USA Everyone is permitted to copy and distribute verbatim copies of this license document, but changing it is not allowed.

0. PREAMBLE

The purpose of this License is to make a manual, textbook, or other functional and useful document "free" in the sense of freedom: to assure everyone the effective freedom to copy and redistribute it, with or without modifying it, either commercially or noncommercially. Secondarily, this License preserves for the author and publisher a way to get credit for their work, while not being considered responsible for modifications made by others. This License is a kind of "copyleft", which means that derivative works of the document must themselves be free in the same sense. It complements the GNU General Public License, which is a copyleft license designed for free software. We have designed this License in order to use it for manuals for free software, because free software needs free documentation: a free program should come with manuals providing the same freedoms that the software does. But this License is not limited to software manuals; it can be used for any textual work, regardless of subject matter or whether it is published as a printed book. We recommend this License principally for works whose purpose is instruction or reference.

1. APPLICABILITY AND DEFINITIONS

This License applies to any manual or other work, in any medium, that contains a notice placed by the copyright holder saying it can be distributed under the terms of this License. Such a notice grants a world-wide, royalty-free license, unlimited in duration, to use that work under the conditions stated herein. The "Document", below, refers to any such manual or work. Any member of the public is a licensee, and is addressed as "you". You accept the license if you copy, modify or distribute the work in a way requiring permission under copyright law. A "Modified Version" of the Document means any work containing the Document or a portion of it, either copied verbatim, or with modifications and/or translated into another language. A "Secondary Section" is a named appendix or a front-matter section of the Document that deals exclusively with the relationship of the publishers or authors of the Document to the Document's overall subject (or to related matters) and contains nothing that could fall directly within that overall subject. (Thus, if the Document is in part a textbook of mathematics, a Secondary Section may not explain any mathematics.) The relationship could be a matter of historical connection with the subject or with related matters, or of legal, commercial, philosophical, ethical or political position regarding them. The "Invariant Sections" are certain Secondary Sections whose titles are designated, as being those of Invariant Sections, in the notice that says that the Document is released under this License. If a section does not fit the above definition of Secondary then it is not allowed to be designated as Invariant. The Document may contain zero Invariant Sections. If the Document does not identify any Invariant Sections then there are none. The "Cover Texts" are certain short passages of text that are listed, as Front-Cover Texts or Back-Cover Texts, in the notice that says that the Document is released under this License. A Front-Cover Text may be at most 5 words, and a Back-Cover Text may be at most 25 words. A "Transparent" copy of the Document means a machine-readable copy, represented in a format whose specification is available to the general public, that is suitable for revising the document straightforwardly with generic text editors or (for images composed of pixels) generic paint programs or (for drawings) some widely available drawing editor, and that is suitable for input to text formatters or for automatic translation to a variety of formats suitable for input to text formatters. A copy made in an otherwise Transparent file format whose markup, or absence of markup, has been arranged to thwart or discourage subsequent modification by readers is not Transparent. An image format is not Transparent if used for any substantial amount of text. A copy that is not "Transparent" is called "Opaque". Examples of suitable formats for Transparent copies include plain ASCII without markup, Texinfo input format, LaTeX input format, SGML or XML using a publicly available DTD, and standard-conforming simple HTML, PostScript or PDF designed for human modification. Examples of transparent image formats include PNG, XCF and JPG. Opaque formats include proprietary formats that can be read and edited only by proprietary word processors, SGML or XML for which the DTD and/or processing tools are not generally available, and the machine-generated HTML, PostScript or PDF produced by some word processors for output purposes only. The "Title Page" means, for a printed book, the title page itself, plus such following pages as are needed to hold, legibly, the material this License requires to appear in the title page. For works in formats which do not have any title page as such, "Title Page" means the text near the most prominent appearance of the work's title, preceding the beginning of the body of the text. A section "Entitled XYZ" means a named subunit of the Document whose title either is precisely XYZ or contains XYZ in parentheses following text that translates XYZ in another language. (Here XYZ stands for a specific section name mentioned below, such as "Acknowledgements", "Dedications", "Endorsements", or "History".) To "Preserve the Title" of such a section when you modify the Document means that it remains a section "Entitled XYZ" according to this definition. The Document may include Warranty Disclaimers next to the notice which states that this License applies to the Document. These Warranty Disclaimers are considered to be included by reference in this License, but only as regards disclaiming warranties: any other implication that these Warranty Disclaimers may have is void and has no effect on the meaning of this License.

2. VERBATIM COPYING

You may copy and distribute the Document in any medium, either commercially or noncommercially, provided that this License, the copyright notices, and the license notice saying this License applies to the Document are reproduced in all copies, and that you add no other conditions whatsoever to those of this License. You may not use technical measures to obstruct or control the reading or further copying of the copies you make or distribute. However, you may accept compensation in exchange for copies. If you distribute a large enough number of copies you must also follow the conditions in section 3. You may also lend copies, under the same conditions stated above, and you may publicly display copies.

3. COPYING IN QUANTITY

If you publish printed copies (or copies in media that commonly have printed covers) of the Document, numbering more than 100, and the Document's license notice requires Cover Texts, you must enclose the copies in covers that carry, clearly and legibly, all these Cover Texts: Front-Cover Texts on the front cover, and Back-Cover Texts on the back cover. Both covers must also clearly and legibly identify you as the publisher of these copies. The front cover must present the full title with all words of the title equally prominent and visible. You may add other material on the covers in addition. Copying with changes limited to the covers, as long as they preserve the title of the Document and satisfy these conditions, can be treated as verbatim copying in other respects. If the required texts for either cover are too voluminous to fit legibly, you should put the first ones listed (as many as fit reasonably) on the actual cover, and continue the rest onto adjacent pages. If you publish or distribute Opaque copies of the Document numbering more than 100, you must either include a machine-readable Transparent copy along with each Opaque copy, or state in or with each Opaque copy a computer-network location from which the general network-using public has access to download using public-standard network protocols a complete Transparent copy of the Document, free of added material. If you use the latter option, you must take reasonably prudent steps, when you begin distribution of Opaque copies in quantity, to ensure that this Transparent copy will remain thus accessible at the stated location until at least one year after the last time you distribute an Opaque copy (directly or through your agents or retailers) of that edition to the public. It is requested, but not required, that you contact the authors of the Document well before redistributing any large number of copies, to give them a chance to provide you with an updated version of the Document.

4. MODIFICATIONS

You may copy and distribute a Modified Version of the Document under the conditions of sections 2 and 3 above, provided that you release the Modified Version under precisely this License, with the Modified Version filling the role of the Document, thus licensing distribution and modification of the Modified Version to whoever possesses a copy of it. In addition, you must do these things in the Modified Version: A. Use in the Title Page (and on the covers, if any) a title distinct from that of the Document, and from those of previous versions (which should, if there were any, be listed in the History section of the Document). You may use the same title as a previous version if the original publisher of that version gives permission. B. List on the Title Page, as authors, one or more persons or entities responsible for authorship of the modifications in the Modified Version, together with at least five of the principal authors of the Document (all of its principal authors, if it has fewer than five), unless they release you from this requirement. C. State on the Title page the name of the publisher of the Modified Version, as the publisher. D. Preserve all the copyright notices of the Document. E. Add an appropriate copyright notice for your modifications adjacent to the other copyright notices. F. Include, immediately after the copyright notices, a license notice giving the public permission to use the Modified Version under the terms of this License, in the form shown in the Addendum below. G. Preserve in that license notice the full lists of Invariant Sections and required Cover Texts given in the Document's license notice. H. Include an unaltered copy of this License. I. Preserve the section Entitled "History", Preserve its Title, and add to it an item stating at least the title, year, new authors, and publisher of the Modified Version as given on the Title Page. If there is no section Entitled "History" in the Document, create one stating the title, year, authors, and publisher of the Document as given on its Title Page, then add an item describing the Modified Version as stated in the previous sentence. J. Preserve the network location, if any, given in the Document for public access to a Transparent copy of the Document, and likewise the network locations given in the Document for previous versions it was based on. These may be placed in the "History" section. You may omit a network location for a work that was published at least four years before the Document itself, or if the original publisher of the version it refers to gives permission. K. For any section Entitled "Acknowledgements" or "Dedications", Preserve the Title of the section, and preserve in the section all the substance and tone of each of the contributor acknowledgements and/or dedications given therein. L. Preserve all the Invariant Sections of the Document, unaltered in their text and in their titles. Section numbers or the equivalent are not considered part of the section titles. M. Delete any section Entitled "Endorsements". Such a section may not be included in the Modified Version. N. Do not retitle any existing section to be Entitled "Endorsements" or to conflict in title with any Invariant Section. O. Preserve any Warranty Disclaimers. If the Modified Version includes new front-matter sections or appendices that qualify as Secondary Sections and contain no material copied from the Document, you may at your option designate some or all of these sections as invariant. To do this, add their titles to the list of Invariant Sections in the Modified Version's license notice. These titles must be distinct from any other section titles. You may add a section Entitled "Endorsements", provided it contains nothing but endorsements of your Modified Version by various parties--for example, statements of peer review or that the text has been approved by an organization as the authoritative definition of a standard. You may add a passage of up to five words as a Front-Cover Text, and a passage of up to 25 words as a Back-Cover Text, to the end of the list of Cover Texts in the Modified Version. Only one passage of Front-Cover Text and one of Back-Cover Text may be added by (or through arrangements made by) any one entity. If the Document already includes a cover text for the same cover, previously added by you or by arrangement made by the same entity you are acting on behalf of, you may not add another; but
you may replace the old one, on explicit permission from t
previous publisher that added the old one. The author(s) a
publisher(s) of the Document do not by this License gi
permission to use their names for publicity for or to assert or imp
endorsement of any Modified Version.

5. COMBINING DOCUMENTS

You may combine the Document with other documents releas
under this License, under the terms defined in section 4 above
modified versions, provided that you include in the combination
of the Invariant Sections of all of the original documen
unmodified, and list them all as Invariant Sections of yo
combined work in its license notice, and that you preserve all th
Warranty Disclaimers. The combined work need only contain o
copy of this License, and multiple identical Invariant Sections m
be replaced with a single copy. If there are multiple Invaria
Sections with the same name but different contents, make t
title of each such section unique by adding at the end of it,
parentheses, the name of the original author or publisher of th
section if known, or else a unique number. Make the sar
adjustment to the section titles in the list of Invariant Sections
the license notice of the combined work. In the combination, y
must combine any sections Entitled "History" in the vario
original documents, forming one section Entitled "Histor
likewise combine any sections Entitled "Acknowledgements", a
any sections Entitled "Dedications". You must delete all sectio
Entitled "Endorsements".

6. COLLECTIONS OF DOCUMENTS

You may make a collection consisting of the Document and oth
documents released under this License, and replace t
individual copies of this License in the various documents with
single copy that is included in the collection, provided that y
follow the rules of this License for verbatim copying of each of t
documents in all other respects. You may extract a sing
document from such a collection, and distribute it individua
under this License, provided you insert a copy of this License ir
the extracted document, and follow this License in all oth
respects regarding verbatim copying of that document.

7. AGGREGATION WITH INDEPENDENT WORKS

A compilation of the Document or its derivatives with oth
separate and independent documents or works, in or on a volur
of a storage or distribution medium, is called an "aggregate" if t
copyright resulting from the compilation is not used to limit t
legal rights of the compilation's users beyond what the individu
works permit. When the Document is included in an aggrega
this License does not apply to the other works in the aggrega
which are not themselves derivative works of the Document. If t
Cover Text requirement of section 3 is applicable to these copi
of the Document, then if the Document is less than one half of t
entire aggregate, the Document's Cover Texts may be placed
covers that bracket the Document within the aggregate, or t
electronic equivalent of covers if the Document is in electro
form. Otherwise they must appear on printed covers that brac
the whole aggregate.

8. TRANSLATION

Translation is considered a kind of modification, so you m
distribute translations of the Document under the terms of secti
4. Replacing Invariant Sections with translations requires spec
permission from their copyright holders, but you may inclu
translations of some or all Invariant Sections in addition to t
original versions of these Invariant Sections. You may include
translation of this License, and all the license notices in t
Document, and any Warranty Disclaimers, provided that you a
include the original English version of this License and t
original versions of those notices and disclaimers. In case o
disagreement between the translation and the original version
this License or a notice or disclaimer, the original version
prevail. If a section in the Document is Entit
"Acknowledgements", "Dedications", or "History", the requireme
(section 4) to Preserve its Title (section 1) will typically requ
changing the actual title.

9. TERMINATION

You may not copy, modify, sublicense, or distribute
Document except as expressly provided for under this Licen
Any other attempt to copy, modify, sublicense or distribute
Document is void, and will automatically terminate your rig
under this License. However, parties who have received copi
or rights, from you under this License will not have their licens
terminated so long as such parties remain in full compliance.

10. FUTURE REVISIONS OF THIS LICENSE

The Free Software Foundation may publish new, revis
versions of the GNU Free Documentation License from time
time. Such new versions will be similar in spirit to the pres
version, but may differ in detail to address new problems
concerns. See http://www.gnu.org/copyleft/. Each version of t
License is given a distinguishing version number. If the Docum
specifies that a particular numbered version of this License
any later version" applies to it, you have the option of follow
the terms and conditions either of that specified version or of a
later version that has been published (not as a draft) by the Fr
Software Foundation. If the Document does not specify a vers
number of this License, you may choose any version e
published (not as a draft) by the Free Software Foundati
ADDENDUM: How to use this License for your documents To u
this License in a document you have written, include a copy of
License in the document and put the following copyright a
license notices just after the title page: Copyright (c) YEAR YO
NAME. Permission is granted to copy, distribute and/or mo
this document under the terms of the GNU Free Documentat
License, Version 1.2 or any later version published by the Fr
Software Foundation; with no Invariant Sections, no Front-Co
Texts, and no Back-Cover Texts. A copy of the license is inclu
in the section entitled "GNU Free Documentation License". If y
have Invariant Sections, Front-Cover Texts and Back-Co
Texts, replace the "with...Texts." line with this: with the Invari
Sections being LIST THEIR TITLES, with the Front-Cover Te
being LIST, and with the Back-Cover Texts being LIST. If y
have Invariant Sections without Cover Texts, or some ot
combination of the three, merge those two alternatives to suit
situation. If your document contains nontrivial examples
program code, we recommend releasing these examples
parallel under your choice of free software license, such as t
GNU General Public License, to permit their use in free softwar

Printed by Books on Demand GmbH, Norderstedt / Germany